TURNED ON

Also available in the Bloomsbury Sigma series:

Spirals in Time by Helen Scales
Chilled by Tom Jackson
A is for Arsenic by Kathryn Harkup
Breaking the Chains of Gravity by Amy Shira Teitel
Suspicious Minds by Rob Brotherton
Herding Hemingway's Cats by Kat Arney
Electronic Dreams by Tom Lean
Sorting the Beef from the Bull by Richard Evershed and Nicola Temple
Death on Earth by Jules Howard
The Tyrannosaur Chronicles by David Hone
Soccermatics by David Sumpter
Big Data by Timandra Harkness
Goldilocks and the Water Bears by Louisa Preston
Science and the City by Laurie Winkless
Bring Back the King by Helen Pilcher
Furry Logic by Matin Durrani and Liz Kalaugher
Built on Bones by Brenna Hassett
My European Family by Karin Bojs
4th Rock from the Sun by Nicky Jenner
Patient H69 by Vanessa Potter
Catching Breath by Kathryn Lougheed
PIG/PORK by Pía Spry-Marqués
The Planet Factory by Elizabeth Tasker
Wonders Beyond Numbers by Johnny Ball
Immune by Catherine Carver
I, Mammal by Liam Drew
Reinventing the Wheel by Bronwen and Francis Percival
Making the Monster by Kathryn Harkup
Best Before by Nicola Temple
Catching Stardust by Natalie Starkey
Seeds of Science by Mark Lynas
Outnumbered by David Sumpter
Eye of the Shoal by Helen Scales
Nodding Off by Alice Gregory
The Science of Sin by Jack Lewis
The Edge of Memory by Patrick Nunn

TURNED ON

SCIENCE, SEX AND ROBOTS

Kate Devlin

BLOOMSBURY SIGMA
LONDON • OXFORD • NEW YORK • NEW DELHI • SYDNEY

BLOOMSBURY SIGMA
Bloomsbury Publishing Plc
50 Bedford Square, London, WC1B 3DP, UK

BLOOMSBURY, BLOOMSBURY SIGMA and the Bloomsbury Sigma logo
are trademarks of Bloomsbury Publishing Plc

First published in the United Kingdom in 2018

A catalogue record for this book is available from the British Library

Library of Congress Cataloguing-in-Publication data has been applied for

ISBN: HB: 978-1-4729-5089-5; TPB: 978-1-4729-5086-4;
eBook: 978-1-4729-5087-1

2 4 6 8 10 9 7 5 3 1

Typeset by Deanta Global Publishing Services, Chennai, India
Printed and bound in Great Britain by CPI Group (UK) Ltd,
Croydon CR0 4YY

Illustrations by Stuart Taylor

Bloomsbury Sigma, Book Forty

To find out more about our authors and books visit www.bloomsbury.com
and sign up for our newsletters

To Caspar, who was literally and literarily
part of the journey

Contents

Introduction

There may be a number of reasons you have picked up this book. Perhaps it was the word 'sex'. Perhaps it was the word 'robots'. Perhaps it was the two words together: a compelling mainstay of science fiction now becoming science fact. Perhaps you judged the book by its cover? It's a nice cover. Or perhaps you bought it as a present for someone to embarrass the hell out of them. Hello, recipient. Perhaps, though, you were curious as to how this could possibly be a subject of any scientific merit whatsoever? That's a perfectly valid starting point and I hope that in the course of the next ten chapters you will allow me to lay out before you the wonderful stories, ideas, science and state of the art technology that show that there is much, much more to a topic that initially seems so trivial.

The past few years have seen a cascade of headlines about sex robots. I've read most of them. In fact, it's often work of mine and my fellow robosexologists (I made that word up just now) behind those headlines. It was never my intention, back in 2015, to become an expert voice for such a niche and somewhat risqué form of technology. However, to absolutely no one's surprise, if you combine the words 'sex' and 'robots' in any form of media, it turns out that people become very animated very quickly. (This is not to say that's what I'm doing *here*, with this book. If you're looking for salacious soundbites in capital letters you're going to be mostly out of luck.)

But this is not a book that's just about sex. Or robots, for that matter. It's about intimacy and technology, computers and psychology. It's about history and archaeology, love and biology. It's about the future, both near and distant: science-fiction utopias and dystopias, loneliness and companionship, law and ethics, privacy and community. Most of all, it's about being human in a world of machines.

★ ★ ★

My own life among the sex robots began, as so many good ideas do, in the pub. I was at a European conference on cognition and robotics and it was full of many different types of researchers working on artificial intelligence. Post-conference drinks are always a great place to pick apart the fundamentals of human existence, especially when the philosophers are there, and if there's one good thing about being friends with philosophers, it's that they share their thoughts on human existence. And if there's one good thing about conferences on cognition, it's that there are plenty of philosophers.

The exact conversation is veiled in a blurry, happy alcoholic haze, but I seem to remember we were discussing the attributes that make us human, that make us feel alive. The conference theme was how we could make systems think: how we might create cognition artificially. This goal – to make machines that can respond to their environment and to situations they've never encountered before – requires knowledge about how we, as humans, do these things. That's not to say we want to emulate a *human* way of perceiving and responding in our machines. That's one way of doing it, but there may be more efficient ways that are better suited to a computer-driven system. But, before we choose a method, first we have to work out how *we* do such things. Even after millions of years of evolution, there's a lot we don't know about how humans work.

Despite our human attempts at sophistication, such as building ourselves nice houses, filling them with tasteful furnishings, dressing ourselves in co-ordinating pieces of clothing and caring about how our hair looks, we can't quite get away from our animal nature. We can pretend all we like that going on a date to the theatre or a wine bar is highbrow and cosmopolitan but it's essentially a mating ritual driven by a need for human connection, maybe (probably) the naked kind of physical connection. Sex is a big part of how humans work. It's why we're all here, millions of years down the line. That brain-fizzing feel-good arousal throws common sense out of the window. It takes a lot to override something so bodily intrinsic to our existence. It affects how we think and how we behave. It affects our perception and our cognition. And it's *fun*.

And so the conversation flowed alongside the drinks and we started to ask questions that didn't have answers; puzzles that we wanted to solve. How, for example, does sex shape the way we think and understand our world, and can – or should – we replicate this in an artificial cognitive system? If a robot is designed to act in a human-like manner, should it be provided with a sexuality? Could we engineer desire? What role is there for sexually active robots in human healthcare? Would this be accepted by society?

I've forgotten most of the papers that were presented at that conference but the questions from the drink-fuelled conversations in the pub afterwards persisted. In the sober light of day, those questions remained pertinent and tantalising. Two events not long after cemented this as a research area I was keen to follow. The first was working with a postgraduate student writing her Masters dissertation on artificial sexuality, a topic I was delighted to supervise. The second was a timely media storm and a call for a ban on sex robots. That was it. I was hooked.

★ ★ ★

Robots have been around us for quite some time. The *idea* of robots has been around for much, much longer – thousands of years, in fact. But it wasn't until the middle of the twentieth century that autonomous robots – machines that could be programmed to carry out tasks by themselves – came into being. These were industrial robots designed to automate factory lines. Much like the steam-powered industrial revolution of the eighteenth century, these robots were the beginning of a more advanced manufacturing process.

We've had three Western industrial revolutions in the modern age. The first was the advent of the steam-powered mills; the second, the use of steel, oil and electrical power. The most recent, the third, is the digital revolution, a product of the Internet and the personal computer. We are now, it is said, on the brink of the fourth industrial revolution: the one where artificial intelligence and robotics disrupt and replace our existing means of production.

Robots and artificial intelligence are two separate things but can be combined to good purpose. The robots are the mechanised bodies. They exist in a physical form – not necessarily humanoid, although that is one form – and they can be instructed to move and respond to programmed orders. Artificial Intelligence, known by its abbreviation AI, is the non-physical part. It's the brains, although 'brain' is a loaded term because AI can't currently think like a human. Instead, it learns from data, analysing the input it has been given, seeking patterns and generating new insights. It's currently pretty limited: there's no artificial *general* intelligence yet, only technology that can perform certain, specific tasks in what we might describe as an intelligent manner. This is not the same as human intelligence. AI can outsmart us playing chess and Go, but we've still got the edge on more reactive reasoning.

Robots and AI are becoming more and more integrated into our everyday lives. AI is all around us. We barely notice it because more often than not it blends in seamlessly. The customer help on that e-commerce website you used earlier? More likely than not, it was a chatbot: automated AI. It's pretty hard to tell the difference between a human working from a customer service script and some software refining its responses. We might not yet have sentient AI – in fact, we might never have sentient AI – but we can certainly make it *seem* human. Put some AI into a robot and that robot can learn about its environment, process that data and act on it in new ways. Give that robot a human form and *voilà*! We're on the way to making an artificial human. Sort of.

A recent phenomenon is the rise in the popularity of virtual assistants: software that can recognise and respond to voice commands. The four of these dominating the market today are Amazon's Alexa, Apple's Siri, Google's Assistant and Microsoft's Cortana. Facebook are said to be releasing their version soon. Speak to one and you can request all sorts of information: the weather forecast, your favourite radio station, recipes and sports results. You can tell them to remember your shopping list or to set an alarm. You can control your lights if you have smart lightbulbs. You can ask them to turn on the heating if your heating is linked to the Internet. You can even ask them to boil your electric kettle, but by the time you've checked there's water in the kettle, refilled the kettle and plugged it in, you might as well flip the switch yourself.

For every order of 'Cortana, lower volume' you can be sure that someone somewhere is making a much more lewd request. Ever talked dirty to Siri? Ever got amorous with Alexa? If you have – even just to see what happens if you do – then you're not alone. If it exists, people will try to corrupt it. But the virtual personalities behind the popular virtual assistants are used to being hit on. In fact, the likes of Microsoft, Google, Apple and Amazon put in plenty of work behind the scenes to rebuff the advances of chatbot-users. Try it. But maybe not at work.

How human is it, though? Enough to trust it? Enough to befriend it? Enough to fall in love with it? The 2013 Spike Jonze film, *Her*, portrayed a near-future scenario where a lonely man going through a divorce falls for his operating system. Alas, the path of true love never runs smooth and the lack of body soon leaves the protagonist frustrated. Fortunately, in real life, we now have a solution for that: enter the sex robot.

Before you get your hopes up, real-life sex robots don't *quite* exist. They've long been a staple of sci-fi but it's only in the past year that they have started to become a reality. The closest anyone has come yet to mainstream commercial production are the creators of RealDoll, California-based Abyss Creations, who have a version of their silicone doll with an animatronic head and an AI personality. Their robot (which isn't quite a robot as it's completely stationary from the neck down) is called Harmony, and you can use their AI app to tweak her personality to highlight the characteristics you admire most. Over in Europe, engineer Sergi Santos has built a robot that needs to be aroused. Your mission, should you choose to accept it, is to give her an orgasm. Notice anything about these examples? Yep, the models being made today are overwhelmingly woman-shaped.

The current versions of sex robots could not be mistaken for a real human. They are something different: cartoon-like, overemphasised and exaggeratedly sexualised portrayals of the female figure. Why, though? If we can make anything we want from the amazing technologies we have at our fingertips, why are we trying – and inevitably failing – to make something realistic? We could engineer whatever we wanted: five breasts, three penises, twenty arms. So why don't we? Is there something in the human form that makes these robots so compelling?

Not everyone is happy about a future of mechanised pleasure. There are legal and ethical issues that need to be ironed out: does sex with a robot count as cheating? Will it lead to violence and rape? What if someone makes a child version? Will it destroy human relationships? Will the robots, as one 2016 headline suggested, 'fuck us all to death'?

Perhaps, though, the opposite is true and the sexual companion robot could instead offer us a chance to enhance our lives: to cure loneliness, to bring us pleasure, to eradicate exploitative sex work, or to treat and rehabilitate sexual offenders. Maybe this is our future, and instead of fearing the rise of the machines we could, quite literally, embrace them.

Let's get stuck in. It's time to explore the fascinating and occasionally murky world of the sex robot. And to begin, we need to travel back at least 30,000 years with a few stops along the way. Here we go. Nothing risqué, nothing gained.

CHAPTER ONE

Been There, Done That

Once upon a time, as all good stories begin, there lived a young woman called Laodamia. Newly married and madly in love, she wept on the shore when her husband left for the Trojan wars. Sadly Protesilaus, her beloved husband, was the first to die at Troy.

> *With sacrifice before the rising morn*
> *Vows have I made by fruitless hope inspired;*
> *And from the infernal Gods, 'mid shade forlorn*
> *Of night, my slaughtered Lord have I required:*
> *Celestial pity I again implore;*
> *Restore him to my sight great Jove, restore!*

The Greek gods, as Greek gods tend to do, let him live again for three hours so that the couple could be together one last time but when the moment came for him to return to the underworld, Laodamia was once more distraught. She was a resourceful woman, however, and she commissioned a bronze likeness of her husband – an artificial lover that she took to her bed. A lovely idea, but a dangerous one. A servant, peering through a crack in the door, saw her kissing, embracing and, as one ancient text says, 'interacting' with the statue and, presuming she was with a man, told her father. He burst into the room (awkward ...), and on seeing the bronze effigy, ordered it to be burned. Oh, it gets worse. Laodamia, unable to bear further grief, threw herself onto the pyre and perished with the figure of her husband. No one lived happily ever after. What we have here is the first written tale of a sex robot, and – in a plotline that goes on to endure for thousands of years – it's a tragic narrative.

The framing of Laodamia's story as a tale about an artificial lover comes from Dr Genevieve Liveley, Senior Lecturer in Classics at the University of Bristol. Within this traditional red-brick setting is a wealth of research into the more prurient side of history. Liveley is the antidote to anyone thinking that classical texts are dull and dry. For years she has been teaching her students about comedy, storytelling, cyborgs, robots and porn – all within the context of the ancient world, and often in Latin or Greek. She and I have been good friends for 18 of those years, from the day I first arrived in England, but it took us 15 years to properly appreciate the overlap in our work.

I forget, sometimes, that not everyone studies such intimate topics. Sitting at a table outside a café in Bristol's ever-so-nice Clifton Village, Liveley and I settle down to slices of cake and cups of tea and discuss all things historically salacious. It is only later, when I receive the outsourced transcribed text of our chat, that I realise someone has spent their morning typing up all manner of ancient depravity, including our digressions, puns and giggles. At one point in the transcript a rather rude word has been substituted with a censorious '[inaudible]'. I'm fairly sure the recording was clear as a bell. I belatedly wonder what the family at the next table thought as we talked about subjects that might make errant eavesdroppers blush. I hope it seemed legitimately highbrow and educational.

In many articles on sex robots, the historical aspect tends to begin with the story of Pygmalion. It's a fairly well-known theme, and one that has been used down the years to popular acclaim, from Shakespeare's *The Winter's Tale* to the eighteenth-century opera *Pigmalion*, the ballet *Coppélia* and the musical and film *My Fair Lady*. On the surface, it's a compelling telling of the sex robot story: boy meets girl; girl is artificial; boy loves artificial girl. But, as Liveley explains, it's much more than that. Indeed, it might not be that at all.

Let's start with the familiar account of Pygmalion, the one we all know from popular culture: a Greek man, a sculptor called Pygmalion, could find nothing good in women so instead he created a beautiful statue. He fell in love with the statue and prayed to Aphrodite, the goddess of love, that he could find a woman like the one he had created. Returning home, he kissed the statue and she came to life. He named her Galatea. They married and had a child.

In fact, Liveley points out, the story is Roman rather than Greek, one of Ovid's making, and is all about deception and delusion. *Metamorphoses*, where we find the story of Pygmalion, is a collection of over 250 classical, traditional myths published in AD 8. There are earlier stories such as Polybius' account of a realistic automaton owned by the Spartan king Nabis – a lifelike robot designed and dressed up to look like his dead wife, Apega. But the tale Ovid tells is his own invention. It focuses first of all on the delusional character of the main protagonist. He fools himself into thinking a statue made of ivory comes to life. 'The whole point,' explains Liveley, 'is that he is deluded. He's a fool, and this is a statue. It doesn't move: he only thinks it does, so this is all in his head, and Ovid takes great pains to emphasise the fact that this is delusional.'

'The other way of looking at it,' she continues, 'is that Pygmalion is either completely deluded, or something wondrous happens and the statue turns into a real girl. So either he's mad and is having sex with an inanimate statue, or something magical does happen and he's actually having sex with flesh and blood. Either way, it's not really comparable to a sex robot. To say it's the story of the first sex robot is misleading.'

Before we scoff at the idea of the character of Pygmalion having sex with an inanimate statue, bear in mind that the notion of agalmatophilia – sexual attraction to a statue, doll or mannequin – was recorded in early Greek civilisation. In a 1975 paper delightfully titled 'Perversions Ancient and Modern', academics A. Scobie and A. J. W. Taylor discuss the classical evidence for statue sex (eleven accounts from Ancient Greece and one from Italy), with the writings of Pliny the Elder telling us: 'There is a story that a man once fell in love with it and hiding by night embraced it and that a stain betrays his lustful act.' Scobie and Taylor's work has been criticised as being unverifiable, although Iwan Bloch, one of the first sexologists, records a paraphilia called '*Venus statuaria*', or 'statue rape'. 'In the case of individuals who are sexually extremely excitable, a walk through a museum containing many statues may suffice to give rise to libido. Of this we have examples,' he writes in his 1909 work, *The Sexual Life of our Time in its Relations to Modern Civilization*.

But, say the scholars, Pygmalion doesn't have that particular peccadillo. Pygmalion's statue was brought to life: he was getting intimate with a living woman (or, indeed, believed that he was). He wanted the real thing, not the representation; it was the living woman he ultimately desired. Do today's sex dolls echo the statues of the agalmatophiliacs? Are people making love to simulacra because they want the living person, or because they are attracted to the sex dolls in their own right? Trudy Barber, an expert in cybersex at the University of Portsmouth, has studied communities where aficionados are sexually attracted to, and in some cases actually aim to become, sex dolls (and sex robots) in a fetish known as *androidism*. 'There is a growing sub-culture,' says Barber, 'of people actually wishing to become robots and dolls explicitly through narcissist forms of sexual arousal and a cult of techno body fascism.'

An essay by media practitioner Allison de Fren explores the world of technofetishism, also known as alt.sex.fetish.robots (ASFR) after the early online Usenet group where the community initially gathered. She observes two groups within this: those who desire an entirely artificial, built robot versus those who desire a transformation from human to robot. Her research revealed that an ASFR ethos was a feminisation of objects: a clear implication and normalisation of gender roles. The crux of it, she writes, is the common interest in programmatic control. The ASFR community wiki describes it as 'a human (typically female) who has been either willingly or unwillingly turned into any kind of inanimate object'. A reverse Pygmalion, as it were.

The centuries-long fascination with the artificial lover endures today. But there is evidence for even older sexual representations of the human body – those that are discreet. And discrete.

* * *

History is the stories of the past that are written down. Before that – before writing – is prehistory. That's the bit that fascinates me. That's the bit where we have to piece together fragments of stories in the same way we do for pieces of pottery: carefully, precisely, delicately; always looking for the most information so that we get it right. There are clues in the materiality of human life – what people made and built and crafted – and in their bodies themselves: the food they ate, the bones they broke, the illnesses they suffered. And, from the get-go, the way that we humans try to make life easier through technology, because technology means using tools, and tools start with a simple stick to poke things with. Speaking of poking things …

In 2005, a 20cm (8in) long, 3cm (just over 1in) wide stone object dated to around 28,000 years ago was found in a cave in Germany. Once it had been carefully reassembled from 14 pieces of siltstone, there was no doubt about it: the etchings and shape make it clear that it was supposed to represent a penis (I am particularly fond of the BBC online news article's excellent use of the word 'tool' in inverted commas*). Described as a 'symbolic representation of male genitalia', it was found in a layer of ash associated with the activities of modern (in an archaeological sense) humans. So was the dildo invented 25,000 years before the wheel?

This is not the first archaeological artificial phallus to be found, but it is one of the oldest. Archaeologists are cautious

* news.bbc.co.uk/2/hi/science/nature/4713323.stm

about attributing a dedicated purpose to them, although this one did show signs of being used as a hammer stone. Indeed, the tendency is often to assume a ceremonial or ritual purpose for such objects. That's not just an embarrassed weaselling-out, though. Down the centuries there certainly are examples of phallic objects being used as charms. The predynastic Egyptian god Min is depicted with a sizeable erection, and phalluses were placed outside homes to honour him. The Indian god Shiva, a deity with prehistoric roots, is also worshipped with the phallic form: the *lingam*. The Parashurameshwara Temple in Andhra Pradesh has a 150cm (5ft) high one dating back to the third century BC. The Romans were also very fond of the penis. It was used as a charm and adorned all sorts of objects – furniture, jewellery, walls of houses. Known as the *fascinus* or *fascinum*, it was said to ward off evil, a sacred invocation of divine protection.

Call me salacious, but I want to know more about the sex-toy side of things. Could our prehistoric counterparts have been using these for more pleasurable purposes? I'm going to go with 'yes'. Maybe not all of them, and maybe not frequently, but when we look at later societies we know this was definitely happening. And if there's one thing that I've learnt from my background in archaeology, it's that human behaviour isn't all *that* changeable.

The use of the word 'toy' in this context is a modern one – a word that reflects the playful purpose of pleasure, unconfined from taboo. I'm using the word to describe its historical manifestations, although it's unlikely they would have been referred to as such. Changing attitudes to sex and sexuality mean we now prefer connotations of frivolity and entertainment, whereas for early sexologists studying the history of these devices, it was a much more serious business.

The Ancient Greeks, as we have discovered a little already, didn't shy away from writing about cavorting. We have the black- and red-figure vases to back up those contemporary written accounts: brushstrokes on pots and plates form figures in all manner of erotic entanglements, many of which include women wielding phallic objects. As far as sex toys go, they did have dildos, which some say could have been made out of bread. Those citizens who wanted a penetrative sex toy could use a hard-baked breadstick known as an *olisbokollix*, literally a bread dildo (plural *olisbokollikes* – still a bit of a mouthful). The validity of this is uncertain: they may just have been referred to by that name because of their shape. There were certainly leather dildos though, and olive oil was used as a lubricant. In Aristophanes' 411 BC comedy, *Lysistrata*, the women of Greece deny their men sex in order to force them to negotiate peace and end the Peloponnesian War, with a wish for an artificial substitute:

And not the slightest glitter of a lover!
And since the Milesians betrayed us, I've not seen
The image of a single upright man
To be a marble consolation to us.

As the Greeks segued into the Romans, Cleopatra – leader of the Ptolemaic kingdom of Egypt and lover of Mark Antony – is rumoured to have invented the vibrator. This vibrator was, apparently, a hollow gourd filled with bees. 'Total nonsense,' retorts British historian Greg Jenner. He calls such rumours 'ficts' – fictional facts that people believe to be true. The truth is more that Cleopatra's reputation as a *femme fatale* was little more than propaganda, a smear campaign by her enemies.

Everything you think you know about vibrators is a phallusy

The origins of the vibrator remain a matter of debate. We've already ruled out Cleopatra as the creator, so who was responsible? One commonly held belief is that the vibrator was invented to treat hysteria in women. This theory may have some merit but it's certainly not definite. We know the word 'hysteria' today as meaning overwrought with laughter, excitement or panic. Prior to the twentieth century, however, hysteria (from the Greek *hysterikos*, meaning 'of the womb') was a fairly common medical diagnosis given to women. This was the notion that a woman's pesky uterus was the cause of all sorts of conditions: nervous ailments, headaches, stomach problems, death … For centuries it was thought that the womb wandered around a woman's body. Literally wandered. Aretaeus of Cappadocia, in the second century AD, described the uterus as being like a living thing inside another living thing. The 'wandering womb' idea persisted right up until the leap forward in anatomical research during the seventeenth-century Enlightenment.

It's been claimed that one way of relieving the symptoms of hysteria was to provide, well, relief. Pelvic massage, to be exact. Rachel Maines' book *The Technology of Orgasm: 'Hysteria', the vibrator, and women's sexual satisfaction* suggests that bringing a woman to orgasm was seen as a cure from the time of Hippocrates onwards. Critics of this theory say that this is a misreading of sources, and Helen King details this at length in her rebuttal *Galen and the widow: Towards a history of therapeutic masturbation in ancient gynaecology*. King is blunt in her disagreement, stating that while it would be nice to have such a neat and continuous narrative for hysteria and masturbation, it may well be another case of Jenner's aforementioned 'ficts'.

Hallie Lieberman, author of *Buzz: The Stimulating History of the Sex Toy* has a PhD in sex toy history. Her research into historical gynaecological texts drew a blank when it came to evidence for using vibrators to treat hysteria. Nevertheless, in the nineteenth century, industrialised nations liked industrialised solutions to whatever ailed them. What we *do* know is that the electromechanical vibrator first appeared in the late nineteenth century, marketed not as a sexual device but as a tool for carrying out vibrotherapy – massage for pain relief in areas other than the genitals, and for men as well as women. In 1883, British physician Joseph Mortimer Granville published a book entitled *Nerve-vibration and excitation as agents in the treatment of functional disorder and organic disease*. You can read this for yourself, it's archived online*, but, with no disrespect to Dr Granville, it's really rather dull. His (lengthy) descriptions read more like those of a TENS pain-relief machine, claiming it will offer relief from neuralgia. And constipation.

Granville does mention hysteria, but claims it is not within his remit and that he wouldn't touch it (or a sufferer of it) with a bargepole, electric or otherwise:

> *In making choice of the several classes of cases in respect to which I now find myself in a position to recommend the use of nerve-vibration, I have been influenced solely by my experience of that remedy. I do not, because I cannot, strongly urge recourse to the method in a considerable number of troublous affections in the treatment of which I have not yet had any large experience of its use. Among these may be mentioned hysteria and the mimetic diseases, and disorders*

---

* ia800304.us.archive.org/31/items/nervevibrationex00gran/nervevibrationex00gran.pdf

> *of the sexual organs, particularly impotence and the affection known, erroneously I believe, as spermatorrhoea**. *I have had cases of the two last-mentioned diseases, some of which have been and are, greatly benefited, one or two apparently cured, but a much larger number of observations than I have yet been able to make, would be necessary to justify any practitioner in holding out a hope that he had actually discovered a new remedy.*

Dr Fern Riddell is an expert on Victorian sexuality. Her book, *The Victorian Guide to Sex*, is a thorough and authentic account of all things sexual in the 1800s, narrated by fictitious characters who bring this to life. Set aside any assumptions of nineteenth-century prudishness: it turns out the Victorians knew a thing or two about getting it on. The take-away message is that the Victorians placed great importance on sexual pleasure within the bounds of marriage. The female orgasm, for example, was seen as vital to fertility and conception.

Masturbation, however, was a touchy subject that was deemed unacceptable and downright dangerous. Founding father of breakfast cereals J. H. Kellogg was a prominent anti-masturbation campaigner, ascribing such activities to all kind of deficiencies, physical and mental. He invented Corn Flakes as a bland diet that would reduce such urges. Having once visited a Corn Flakes factory, I can confirm that it was one of the least arousing places I have ever been. The white coats and blue hairnets might appeal to production-line fetishists, but the thick rubber gloves would get in the way of acting on this.

* Spontaneous ejaculation – probably referring to wet dreams.

So how did the vibrator emerge as a sex toy? It's not quite clear. The dildo was still around, unmoving – that much is clear from texts and illustrations. But early vibrators were in no way being openly marketed as a sexual device. When they weren't being used as a treatment for sore muscles, vibrators were being sold as beauty devices. There's every probability that they were being put to other uses; it's just that there isn't really a concrete record of it. Early twentieth-century advertisements alluded to their potential but couldn't be forthright about it. Magazines contained lines like 'continued use over different nerve centres will bring an undeniable tingle which has not been felt in a long while'. Laws and propriety forbade any sexual references and so the explicit was made implicit. For the female body, where orgasms predominantly (if not entirely) depend on clitoral stimulation, this was a wonderful breakthrough.

In 1953, a publication by US sexologist Alfred Charles Kinsey on the sexual practices of over 6,000 women reported vibrator use as 'rare' – used by less than 1 per cent of those surveyed. Even into the 1970s, companies were reluctant to advertise vibrators as anything other than generic health devices. At least they were being advertised; dildos, by contrast, were unambiguously intended for sexual purposes and so were very much hidden from public view. In 1968, Japanese technology company Hitachi started selling their 'Hitachi Magic Wand Household Electric Massager' in the US. It was a massive hit: a very obviously sexual one.

At the time of writing this, Betty Dodson is 88 years old. Conditioned as we are to think of sex as the domain of the young and passionate, it may be surprising to find that Dodson is responsible for some of the most fundamental changes in attitude to women's sexual pleasure. And she's still at it – 'it' being both education and sex itself. She is open and frank about everything from how to have the best orgasms to the joys of sex with her much younger lover (he was 22 when she was nearly 70). Dodson, in my opinion, is an absolute legend.

Dodson refers to herself as a 'pro-sex feminist' and is heralded as being one of the key players in the US women's liberation movement. In the 1970s and 1980s, Dodson used the Magic Wand as a teaching tool, holding masturbation workshops in her apartment in New York. She discovered that most of the women she met had been discouraged from exploring their sexuality, and many had never experienced an orgasm. Her solution was to host classes of 10 to 15 women and coach them in learning about their bodies and their sexual responses. In 1973, she selected the Magic Wand as the tool of choice, after two years of experimenting with different vibrators. She was never paid to endorse it but, thanks to her, it became the most popular vibrator of all time.

Even though it had gained a reputation as a sex toy, in 1999 Hitachi were still maintaining their popular product wasn't for masturbatory purposes. Then, in 2013, they decided to stop manufacturing it owing to concerns about the company name being linked to sex. Quite why it had taken them so long to notice this is unclear, especially given that it had featured in pornographic videos for quite some time. Fortunately, rebranding was the solution, and the vibrator was renamed the Original Magic Wand to omit the Hitachi name, even though they were still manufacturing the product.

The opening of sex shops in the 1970s meant that sex toys with an undisguised sexual purpose could be sold (though in some places laws forbade – and sometimes still forbid – their sale. Hard luck, Alabama). They remained outside the mainstream, though. Purchasing a sex toy meant braving a niche shop or relying on mail order. The 1980s saw the rise of home parties, such as Ann Summers parties in the UK, where women could view and buy sex toys in an atmosphere deemed to be fun. Sex toys were being bought, but it was still happening behind closed doors.

In 1998, the writers of HBO's television series *Sex and the City* wanted to include the use of a vibrator in the storyline for their Season One episode, 'The Turtle and the Hare'. The model they chose was a rabbit vibrator: a dual-action sex toy with a rotating shaft and two small external 'ears' that vibrate against the clitoris. The rabbit design came from Japan, where obscenity laws led to the abstraction of phallic designs into something bright, cute and colourful. When the episode aired, sales rocketed. Here was the depiction of sexual pleasure with a toy that optimised the penis. Better still, the advent of online shopping meant that it could be ordered quickly and delivered discreetly to your door. Everyone was talking about it. The media were covering it. Sex shop Ann Summers sold one million rabbits in a year.

In terms of respectability, however, Hallie Lieberman is rightly critical of the attitudes depicted in the *Sex and the City* episode. She points out that while the show's inclusion of sex toys and masturbation might seem progressive, the storyline was judgemental. Charlotte becomes addicted to her vibrator and sinks into isolation, scaring her friends, who intervene to 'save' her from herself. It's not, she notes, the only episode where a vibrator is portrayed as a poor substitute for a man, and where masturbation is shown as an inferior form of sexual pleasure.

Nonetheless, people began to talk more openly about owning sex toys, although the jokey approach was still dominant. Twenty years on, the landscape has shifted.

These days, the sex technology industry is forecast to generate profits worth £22 billion ($30 billion) worldwide by 2020. It ranges from hardware (like sex toys) to software (such as hook-up apps that allow you to find sexual partners with just a swipe of a phone screen) and new forms of content delivery, where technology such as virtual reality is used to provide a tailored sexual experience. That figure of $30 billion doesn't include online porn, which alone generates as much as the Hollywood movie industry.

And the penetrable sex toy, the *cunnus succedaneus*, or substitute vulva? Vibrators were – and still are – used by men. Likewise, penetrative toys such as buttplugs or anal beads to pleasurably stimulate the prostate are popular. Representations of the vulva and vagina are comparatively commonplace throughout time. But as a sex toy that can be penetrated? There isn't as much in the way of material evidence. Or, at least, not in a form that has survived in the archaeological record. In some ways, we could attribute this to the lack of need of sex toys for men. Their sexual desires were often catered to by someone else, after all. Where there was a distinct lack of partners available, a hand was an easy solution. Yes, there's a suggestion that substitutions were made in the form of fabric or leather, mimicking orifices. Alas, these materials don't survive well. Nor do their rubber counterparts.

More modern examples do exist, as do references to them. They are the subject of Cynthia Ann Moya's fascinating doctoral thesis (2006), written at the Institute for Advanced Study of Human Sexuality. She recounts an advertisement in a French catalogue from 1902, translated by writer Henry Nathaniel Cary. It describes in glowing terms the charms of an inflatable woman's belly and pelvis that can be 'folded up and placed in the pocket as easily as a handkerchief'.

A 1927 book shows a photograph of a stuffed artificial vagina, much like a cushion with a central cavity, surrounded by fake pubic hair. Moya points out that pubic hair was also a feature on many early dildos, although it is generally missing from sex toys today. Times and tastes change.

Kinsey's 1948 report on male sexuality, *Sexual Behavior in the Human Male*, claims that 'sometimes devices which simulate the female genitalia may be used for masturbation, but they are rarely employed'. Men, he tells us, are largely manual, although his surveys found that some men made 'rare to occasional use of bottles, tubes, holes in objects'.

A 1940s sex catalogue from Japan shows a rubber pocket that can be inflated, heated with warm water and then deflated again for ease of portability. Such artificial vaginas are acknowledged in older Japanese folklore. Known as *azumagata*, they are depicted in nineteenth-century *shunga* (Japanese erotic art) and in eighteenth-century texts.

But it wasn't until the mid-1990s that the portable artificial vagina became a commercial reality in the Western world. American Steve Shubin was facing months without sex after his pregnant wife was told to refrain from intercourse due to a high-risk pregnancy. Shubin's foray into sex shops (with his wife's blessing) left him disheartened so he decided to take matters into his own hands. He invented the Fleshlight®, a masturbatory device shaped like a hollowed-out flashlight, with a name now used erroneously to describe most similar types of devices. It was quite the hit.

Today, Japan is one of the leading manufacturers and largest markets for male sex toys. Japanese company Tenga, a self-proclaimed 'lifestyle brand of sexual wellness items', has sold over 57 million penetrable sex toys globally over the last 13 years. They are very clear that their products are not designed to replicate human anatomy, rather that they are to enhance masturbation. Their products range from textured, one-use sleeves to electronic masturbatory devices. None of them resemble body parts: instead they are modern, almost space-age, white, soft sculptures.

As a little experiment, and because I get sent a lot of sex toys for work, I decided to share some out among my male friends, so I sent some of them one of Tenga's Eggs each. The Egg is a stretchy thermoplastic elastomer (TPE) sleeve that comes packaged in a hollow plastic egg that fits in the palm of the hand. Each Egg also contains a sachet of lubricant. The sleeve itself has embossed designs on the inside: ribs, spirals, that kind of thing. It fits over the head of the penis and is held in place as the man masturbates.

The reaction from my friends was mixed. Friend One, who was horrified at the idea, declared: 'I am *not* putting my dick in that.' Friend Two was ambivalent: 'It's okay but I wouldn't bother paying for one. It's essentially just a posh wank.' Friend Three declared 'my hand is better'. Friend Four was willing to offer feedback but I really wasn't happy hearing about it in detail because I've known him since he was born; he acknowledged that it was 'okay'. Friend Five said 'meh'. My friends, it seems, are in agreement with Kinsey that 'most males restrict themselves to a limited series of particular techniques to which they have been erotically conditioned'.

In a 2013 Europe-wide survey, Internet-based market research and data analytics firm YouGov found that 33 per cent of 2,168 UK respondents had used a sex toy. British company Lovehoney sell a product every 16 seconds. The world's favourite sex toy is a dildo, they say, closely followed by cock rings and vibrators in their many forms. Penetrable toys don't make the top 10 at all.

The past few years have seen new start-ups come onto the scene, capitalising on advances in materials and technology and an appetite for pleasure. The rabbit vibrator was a change in design: a move away from replicating human body parts. It showed that the form of sex toys could change, could be abstracted, could be optimised. Today, there are hundreds of vibrators, dildos and fleshlights available, many of which look nothing like a human body part. Some are unusual and sinuous shapes that seem beautiful and ornamental. There are vibrators that curve and that bend and those that are malleable. There are artificial orifices in alien form. They can be programmed with different pulsing patterns for the user to select. These new products have another advantage: they're smart.

Teledilwhatnow?

Trudy Barber has been researching intimacy and technology for a long time. As an undergraduate student at Central Saint Martins art school in London she developed one of the first immersive VR sex environments in 1992 (and, she proudly states, the first erect male penis in virtual space). She went on to complete a PhD in Computer Fetishism and Sexual Futurology, a particularly avant-garde subject even now. She's smart, funny, bright and inspiring. Her boundless energy and passion is contagious: when Barber starts to talk, everyone stops to listen. She's a pioneer of teledildonics and the queen of cybersex – a way of exploring how people use various types of technology for sexual purposes. She is visionary, and I adore her.

The word teledildonics comes from combining the Greek word *tele*, meaning remote, with a reference to dildos. In other words, it's a form of technology where sex toys can be controlled remotely, allowing the user with the sex toy to feel tactile sensations that have been initiated by someone else, somewhere else. In the first wave of virtual reality (VR), teledildonics was going to revolutionise our relationships. The promise was new sexual sensations and remote pleasure. Alas, the first wave of VR was limited by cumbersome equipment, slow data transfer and the decidedly unsexy nauseating effect of motion sickness. But just because it wasn't quite happening didn't mean developers were deterred.

Barber's work has taken her all over the world to events both public and private. She regales us with tales of walking into a room to give a talk to a group of Silicon Valley tech bros in the early days of VR, and leaving them all flustered and blushing when she removed her dress to reveal her sex suit underneath (it was really just a designer body stocking with a corset but the effect was perfect). As I watch early footage of her I am struck by how resonant it remains today. In a 1993 video she enthuses about the potential of VR to let people experience other bodies, including gender swaps, something that became a global success for BeAnotherLab in 2014 with their project, 'The Machine to be Another', which allowed a user to convincingly feel as though they were inhabiting someone else's body.

Twenty-five years ago, Barber was musing on the cult of celebrity: the idea of replicated identities and a need for copyright and licensing. This is currently big news with the recent advent of Deep Fakes, where artificial intelligence has been used to seamlessly swap celebrities' (and non-celebrities') faces with those of porn performers, creating footage of sex that stars someone who was definitely not involved with making it.

Barber is optimistic, seeing great potential for happiness. 'I try not to get too serious about this, because it's so funny,' she smiles. And she's right. Humour is a glorious part of this, hand-in-hand with pleasure. Her current work is taking her back into the new wave of VR, exploring more about the body and uses of virtual space. 'We're dealing with human nature,' she reminds us, 'and basically it's kind of same-old, same-old, but with different technology.'

★ ★ ★

I'm sitting on the steps in the park opposite the San Francisco Museum of Modern Art. It's the rendezvous point for meeting a man who engineers smart sex toys for a living. Kyle Machulis, also known as qDot, has been building teledildonics – sex toys synced over the Internet – for well over a decade now. Next to him, I'm an amateur. My own knowledge of the subject is completely dwarfed by the amount that Machulis knows. He's Mr Sextech.

Machulis arrives bang on time. A clean-cut guy in glasses and a tech T-shirt (his day job is coding for Mozilla), he doesn't look as if he's spent the past 14 years immersed in sexual subcultures. But then, it's the quiet ones you've got to watch. Having talked to him at length online, I'm delighted to meet him in person. We walk to a nearby coffee shop and grab the last free table, and I begin by asking him about his work.

Smart sex toys are becoming more and more standard. They don't need to be used via VR: they can simply be connected to a smartphone via an app. That phone is, in turn, connected to the Internet and so the toys can also be controlled online. For example, a sole user could control the settings of a vibrator via their phone app, or could share control over the Internet so that someone else could join in and interact remotely. There are all sorts of smart sex toys available, including those that can be paired so that two (or more) people can have a stimulating experience with each other even if they are in different places.

Machulis runs the website Metafetish, a site 'about sex toys, sex technology, and other sex stuff. Whatever.' Metafetish is a great resource for anyone building or hacking sex toys. Its chat forum is populated by intriguing questions such as 'Weirdness with Fleshlight Launch command interruption?' and 'Was the buttshock arduino shield just a phase?' (Answer: probably.)

Machulis has archived a wealth of resources around teledildonics, from early media, including digitised copies of 1990s San Francisco erotic magazine *Future Sex*, to his own work, which started around 2000. His work began by hooking up a sex toy to the video game *Quake*, a first-person shooter game. As a bit of a joke he decided to wire things up so that shooting characters made a vibrator go faster.

Next came SeXbox, a 2005 open-source sex platform made from an Xbox controller and a vibrator. This was the start of Machulis' interest in democratising the sex tech scene. His goal of DIY and open-source equipment gives control to the user – something that he would later build on with his project Buttplug,* a 'set of specifications and applications that allows users of sex toys to be able to use the toy they like with the applications they want'.

Machulis' aim of levelling the playing field of teledildonics is a noble one when you consider how this kind of technology is currently used. Many of the smart sex toys available today are used in sex work. Webcam models perform sex acts

* The name 'Buttplug' does not mean the only sex toys that can be controlled are buttplugs. Multiple types of sex toys are supported. Machulis picked that name because it's a toy that can be used by anyone.

online for money, streaming the videos over the Internet in a live broadcast. These sex workers can use devices linked to those of the paying viewer to provide them with a remote sexual experience. This often takes the form of a vibrator paired remotely with a penetrable masturbation device. The usage patterns of the cam model's toy can be synced with the viewer's toy, allowing the viewer to feel similar motions and vibrations, making the act seem more mutual and connected. But cam models are reliant on proprietary software to control their devices. These can have costly licensing fees. Machulis wants to remove the limitations imposed by the device manufacturers, allowing the workers to take control – a kind of X-rated socialism.

The world of sex tech is an exciting one. It does have its limitations, not least regarding funding. Venture capitalists are often prohibited from investing in anything deemed 'adult', so investment is hard to come by. The rise of hacker communities and maker spaces, continuing the work of Machulis and others, is therefore an important part of innovating new products. Sex toys, as discrete body parts, have been around for millennia, and are as popular as ever today. But what about something bigger, something more … human?

Guys and dolls

Enter the sex doll. The use of a sex doll suggests something more than the use of a smaller, disconnected sex toy, yet the intrinsic actions are pretty much the same. An easily ignored small sex toy like a vibrator is a handy way to compartmentalise thinking about sex. Hidden in a bedside drawer, it is out of sight and out of mind when not in use. Not so the sex doll: it takes up space and is an entity.

It wasn't until the latter part of the twentieth century that sex dolls became widely available commercially, but they have a much longer history. There are written references to seventeenth-century sailors creating *dames de voyage*, or *dama de viaje*: women-shaped bundles of fabric and leather for sex-starved sailors to share. In fact, Japan has had soft, cushioned, fabric sex dolls called *datch waifu* – Dutch wives – a reference to the dolls on the Dutch ships they encountered in trading.

The blow-up doll first made an appearance in the 1970s. Often sold now as a novelty joke item, it was actually banned in Britain until the late 1980s on the basis that it was obscene. Now a staple of comedy programmes and stag nights, its origins as a sex aid have been muted somewhat. With its O-shaped mouth and rudimentary hole for a vagina, it is a comic-book caricature of a woman. There are inflatable male dolls too, though these seem to have been a much later development to capitalise on the novelty market.

Dolls, as we have seen, have been venerated for quite some time. Fetishised, too, in the manner of the aforementioned Pygmalion, or the agalmatophile. It wasn't until the mid-1990s that the realistic sex doll became a more public commercial reality. Today, they are fairly widely available, though the market is niche. The cheapest, costing under £750 ($1,000), are the 'mini dolls', which are easy to store, say the advertisements, and good for trying out before buying a full-size version. Nonetheless, their child-like size is incredibly disturbing. The mid-range examples, costing around £2,200 ($3,000), are made from inferior materials compared with the high-end dolls, but are reasonably similar in terms of appearance until you get up close. High-end dolls are handcrafted using quality materials.

You'll pay anywhere from £3,500 to £7,000 ($5,000 to $10,000) for one of these.

It won't have escaped your attention that, by and large, all the sex dolls available today take a female form. There are some male versions, but they represent a tiny share of the market. There has been plenty of speculation over why this might be. 'Women don't want them' is a common response. We'll look at that in more detail later, because this particular aspect of sex dolls – and, by extension, sex robots – warrants a lot more attention.

It's difficult to pin down statistics on who is making sex dolls, where, and how many are sold. There is certainly evidence of Chinese factories: Guangzhou New Sino Environment Technology Company is one of these, making their TPE Hitdoll and silicone Z-onedoll. Doll Sweet is another multinational company operating out of China. There are Japanese manufacturers too: Orient Industries (who make CandyGirls) and 4Woods (with their A.I. Doll and Naughty Dolls). Russia has a small workshop, Anatomical Doll. In Europe, German company Mechadoll and French company DreamDoll are the main manufacturers.

In the US, companies like Private Island Beauties and Silicone Wives vie for attention alongside probably the most well-known creators, Abyss Creations, who make RealDolls, 'The World's Finest Love Dolls!', the state of the art in artificial lovers. Abyss Creations have been producing RealDolls since 1996. The dolls have a poseable PVC skeleton with steel joints and silicone flesh, customisable genitals, swappable faces and a menu of body shapes. You can choose the features from a range of options to suit your specification, provided you have upwards of £3,500 ($5,000) to pay for it all. RealDolls have become a byword for sex dolls, for the most part due to the publicity they have received over the years, including a starring role in the 2007 film *Lars and the Real Girl*. They have competition in the shape of Los Angeles-based firm Sinthetics, a husband-and-wife team who run a company making what they prefer to call 'manikins'. Like Abyss Creations, they also specialise in beautifully handcrafted products.

A few years ago, Abyss Creations announced they were working on a sex robot with their spin-off company, Realbotix, spurred on by requests for interactivity in their dolls. Any month now (although they've said 'any month now' quite a few times) they will release their first commercially available sex robot.

But what *is* a robot? What do we mean when use that term? That, my friends, is a whole other chapter waiting to be unpicked.

CHAPTER TWO

I, For One, Welcome Our New Robot Overlords

Technology in its broadest sense refers to the use of tools that can solve real-world problems. These aren't just digital. The earliest examples we have are stone tools from 3.3 million years ago, probably used to butcher animals. These tools, examples of which have been found in Kenya, were being created half a million years before the arrival of the genus *Homo*, the biological classification that includes us, the species *Homo sapiens*. The tools are crude. The untrained eye might not have even noticed them: they look to all intents and purposes like some fractured rocks. But they represent a major step in technological development. These rocks weren't just being picked up and used to hit things. There was deliberate reshaping of the stone, including sharpened edges. Our predecessors *Australopithecus afarensis* were making their lives easier by creating basic technology.

Ask a child to draw a robot for you and they will probably produce a sketch of a small box on top of a larger box, with rigid arms and legs protruding at stiff angles and clunky feet and hands in a fixed grip. Television of the 1960s has become engrained in our cultural memories, even for those born decades later. This instantly recognisable portrayal of a robot – a literal icon to the point where the robot emoji takes exactly that form – means that we often don't describe the robots around us as actually being robots.*

Ask someone to define 'robot' and the answer might be a little vague. It's hard to pin down. When does a piece of equipment stop being merely a piece of equipment and start being a robot? The International Organization for Standardization, in the publication *ISO 8373:2012 Robots and robotic devices* (catchy, eh?), has a very specific definition: 'actuated mechanism programmable in two or more axes [direction used to specify the robot motion] with a degree of autonomy [ability to perform intended tasks based on current state and sensing, without human intervention], moving within its environment, to perform intended tasks'.

In other words, robots have actuators (moving parts and parts that move them), a control structure (so they can be told what to do), sensors (to make sense of the surrounding environment) and therefore an ability to do tasks on their

---

* In order to test this, I asked my seven-year-old daughter to draw a robot. She said: 'Is a protocol droid okay or do you want something in particular?' I had to ask another child instead.

own. They are programmable machines designed to automate a task in a series of fairly complex actions. They do the jobs that robotics communities term the '4Ds': dumb, dull, dirty and dangerous. We see them at use in the home and garden (for example, robot lawnmowers), in hospitals (robotic surgery), in the military (for bomb disposal, or for automated weapons systems), or in farming (such as robot milking machines).

These days, because robots are no longer confined to production lines, we refer to them as *situated*: they are part of an everyday environment that rapidly changes around them, and they can respond to that, and detect and act on those changes. Their presence in that physical environment means they have a body, through which they experience new information. That body might be just a box with cameras or an arm with sensors, but it's how the computer brain of a robot enacts with its surroundings. We humans are embodied, too: we understand the world around us through our perceptive senses – vision, touch, smell, taste and hearing. Robots are similar: whether it's a robot vacuum cleaner that uses computer vision to clean your floor, or the Curiosity rover on Mars, sampling climate and geology, the physical form of the robot is in some way dictated by the physical interactions it needs to have.

Robomyth and robotruth

We've always been dreaming up new technology that can assist and benefit our lives. From hitting things with stones to delivering things with drones, we do love a chance to reduce our labour. One of the easiest ways to imagine all our jobs being done for us is to imagine a functional version of ourselves. No wonder the idea of artificial humans is so popular. The precursor of the robot, the automaton (plural automata), has a lengthy history. Automata are machines that give the appearance of being self-powered and self-driven, working independently. The name comes from the Greek, meaning 'acting of one's own will'. In fact, they are mechanistic, performing repetitive pre-set actions that might seem self-selected but are merely automated.

Homeric myths tell of the Greek god of fire, Hephaestus, who was said to have the power to produce motion. Hephaestus was a fairly prolific blacksmithing deity. As well as Eros' bow and arrows, Hermes' winged helmet and sandals and Aphrodite's girdle, Hephaestus created 20 bronze-wheeled self-propelled tripods. These automata were used as servants, trundling in and out of the halls of the gods. He also built bronze watchdogs, fire-breathing horses and the giant eagle sent to torture Prometheus. He created human-like figures, too: the giant Talos, protector of Crete, a Terminator-like figure who patrolled the island, warding off pirates by throwing rocks at them. In other stories, Talos is described as 'the last of the bronze generation of men'.

Story it may be, but in real life, in 1972, two full-size Greek bronze castings of naked men were found in the sea at Riace in Calabria, Italy. The bronzes, dating from 460–450 BC, are housed in the Museo Nazionale della Magna Grecia in Reggio Calabria. Their teeth are silver; their lips and nipples, copper. Standing around two metres (6ft 6in) tall, these early classical statues bring to mind the myths of the godly smith and Laodamia's recreated husband. The Riace bronzes are no automata, but they are a fine display of the skill of the Ancient Greeks in creating lifelike sculptures.

There are examples of real automata from the Hellenistic period in Greece, though. The most famous of these is the Antikythera mechanism – an incredibly complex clockwork analogue computer for predicting astronomical positions, made mainly of bronze and created sometime between 250 BC and 60 BC, discovered on a shipwreck off the coast of the Greek island of Antikythera in 1902. The sheer quality and sophistication of this mechanism is testament to the skilled scientific accomplishment of the time. It is intricate and fascinating, and there are no parallels to its complexity until mechanical astronomical clocks were developed in Europe some 1,500 years later. I am in awe of the Antikythera mechanism, not just because it's a 2,000-year-old computer of astounding ingenuity, but also because if it was possible to make that, then what else were the Greeks making?

In Rome and Egypt, it's reported that moving statues – puppets, more or less – made gestures and transmitted wisdom. Strings and levers were the mechanisms behind these. John Cohen, in his 1975 book *Human Robots in Myth and Science*, describes these in a way that brings to mind *The Wizard of Oz*, a film that draws on millennia of tales about oracular heads dispensing wisdom and orders. The speaking head pops up a lot: stories abound about brazen heads – mechanical heads made of bronze that could speak and prophesise.

E. R. Truitt is Associate Professor of History at Bryn Mawr, and she's an expert in mediaeval robots, both real and reported. Her book *Medieval Robots* describes how, in the Middle Ages, accounts of fantastical automata appeared in manuscripts across Europe. Robots were a form of wonder and entertainment before they became a reality.

A throne belonging to King Solomon is mentioned in the Bible, in the first Book of Kings, which was written around 550 BC: 'Moreover the king made a great throne of ivory, and overlaid it with the best gold. The throne had six steps, and the top of the throne was round behind: and there were stays on either side on the place of the seat, and two lions stood beside the stays' (1 Kings 10:18–19). Then, 1,500 years later, a much more elaborate version of this marvel was reported at the Byzantine court at Constantinople. This Throne of Solomon, it's said, was a gold-covered, ivory wonder bedecked with rubies, sapphires, emeralds and mechanised animal figures that carried the ruler up the steps and held his heavy crown. It was described by Byzantine texts and by the tenth-century diplomat Liudprand of Cremona. So was there a real Throne of Solomon, or was it the stuff of myth? There's no archaeological evidence for it but it's possible that the Byzantine emperors used the legend of Solomon's throne to build their own mechanical version, though just how elaborate that version was is up for debate.

There were tales of automata from across the globe. Ancient Baghdad had moving statues powered by the wind, and mechanical birds singing on artificial trees. Ming Dynasty China apparently had tiger automata. In 1495, Leonardo da Vinci sketched the design of a robot knight that could move its limbs. A twentieth-century version of this was made to his design and it worked.

Truitt's book describes a wealth of medieval automata, and examines the concepts behind them. In one page from a fifteenth-century illuminated manuscript of *Le Roman de la Rose* (*Romance of the Rose*), a woman is shown working at a forge, creating people out of body parts, with human hands and feet strewn on the floor around her.* This is

---

* Delightfully, Truitt has also posted on her blog a photo of a fifteenth-century pewter badge, which shows a penis as the body part of choice: 'a naked Natura stands at her forge, whanging away at a wang'. (www.medievalrobots.org/2012/10/natura-artifex-redux.html)

Natura the artisan, a metaphor and representation of Nature. It was believed that God created nature, but humans could only copy it, making mechanical counterfeits. Automata are mimetic. They could definitely be impressive, but they were viewed as mere simulacra when pitted against the wonders of God's creations.

In the late Middle Ages and early Renaissance, astronomers were explaining the universe in terms of clockwork: *machina mundi*, the machine of the world – a clock wound by God. The cosmos, and the creatures living within it, was envisaged as a series of precise, moving parts. This belief in the harmonious movement of the skies and the systems governing our bodies was echoed in the creation of automata. In France in the fourteenth century the Count of Artois furnished his garden with mechanical monkeys covered in fur to trick visitors into thinking he was important enough to have real ones. The creation of automata grew throughout the fifteenth century, indicating wealth and power in those who owned them.

The clockwork monk is a beautiful example. These 38cm (15in) tall figurines with a clothed wooden body and an internal clockwork mechanism could be wound up and set to walk across a surface while apparently praying and kissing their rosary beads. Showcased in the London Science Museum's *Robots* exhibition, these automata, which are nearly 500 years old, are a fascinating glimpse of religious statuary brought to life. In X-ray pictures the chains and cogs form the metal skeleton and muscles, their actions in sync like the heavens. Around the same time, *zashiki karakuri* were being made in Japan – dolls with hidden mechanisms that could serve tea and paint using brushstrokes.

The late sixteenth century saw the myth of the golem gaining more traction. In Jewish folklore, a golem was a clay figure of a person animated by magical means. It was said to be brought to life by letters – specifically one of the hidden names of God in Hebrew – whereupon it could serve as a worker. In a way the creator was the programmer, actually entering code into the hardware. The details of the legends bear relevance to robots today: a golem was obedient to its owner but performed tasks literally, without any sentience. Those tales, like much of our robot sci-fi, didn't always end well. Many of the myths centred on the golem going rogue.

By the seventeenth century, automata were becoming more sophisticated and more impressive. One of the most famous creators of these was Jacques de Vaucanson, a skilled toy-maker. In 1738 he built a flute-playing automaton – a life-size wooden figure containing a system of bellows that allowed it to 'breathe' and play the flute. This was no mean feat: the flute is a difficult instrument to play and requires the controlled and careful movement of tongue, breath and fingers. Through meticulous crafting, Vaucanson created an automaton that could independently play twelve tunes. The sophistication was ground-breaking.

Vaucanson's next ambition was to create an automaton that gave the impression of visceral responses: eating, drinking, digesting and excreting. He did this in 1739 by building his famous masterpiece, the Digesting Duck. The size of a real duck, it had over 400 moving parts. It didn't *actually* eat and digest and defecate – Vaucanson faked that – but it could seemingly peck at food and water and swallow it, and then excrete dyed breadcrumbs that had been put in place beforehand.

Vaucanson had raised the bar. Thirty years later, Hungarian inventor Wolfgang von Kempelen produced his famed Mechanical Turk, a chess-playing machine in the form of a human-like torso and head behind a cabinet on which was displayed a chess board. It could nod or shake its head at moves made by its opponent, could grasp the pieces and move them, and could even pull faces. It was also very, very good at chess. Alas, it was fake. The cabinet contained an intricate set-up of gears and cogs but these were only in place to fool anyone examining the would-be automaton. The reality was that the cabinet hid a master chess player who was controlling the mechanisms from the inside. Although it was subject to many sceptical attempts to reveal its true nature, it wasn't until 1857, nearly 100 years later, that the full extent of the secret workings was revealed. Nonetheless, the Mechanical Turk was a hoax that raised fundamental questions around automation and machines that not only perform labour but can, perhaps, think.

The Industrial Revolution, which began in the late eighteenth century, had an enormous impact on how people thought about the automation of labour. Britain paved the way in industrialisation, introducing mechanisation, steam power and machined tools. The results were dramatic, leading to innovation across vast swathes of everyday life. The impact was not always positive: although the standard of living overall increased, in specific areas life became much harder, particularly regarding child labour and health. Advances in mechanisation led to more groundwork in automation. The second half of the nineteenth century saw the increased popularity of automata, when the scale of production grew and they became commercially attractive.

But the robot? That had to wait until the twentieth century.

Forced labour

The word 'robot' has its roots in an eastern European term for servitude. It comes from the Czech *robotnik*, meaning 'forced worker', and from *robota*, which describes drudgery and forced labour. The word 'robot' was specifically used by the Czech playwright Karel Čapek in 1922 in his work *R.U.R.* (*Rossum's Universal Robots*) to describe human-like manufactured people. He attributes his use of the phrase to his brother, the writer and artist Josef Čapek, who coined it for a short story. Karel's envisioned robots were made from synthetic organic parts, more in line with *Blade Runner*'s replicants than with machines. They were, in essence, human clones. The name stuck, though. In Čapek's play, the robots are incapable of original thinking but, as in all good dystopian fictions, some of them gain self-awareness and rise up against their human creators. Sounds familiar? It's a sci-fi mainstay.

Most of the specialised robots in the world today are used in manufacturing, particularly in automotive industries. China is by far the market leader. A 2017 report by the International Federation for Robotics put the figure at 1.8 million industrial robots in current use in factories worldwide. A further 285,000 are thought to be currently in action as service robots for professional use – things like defence robots, medical robots, underwater robots and agricultural robots. Mass-market domestic and personal robots, however, are much cheaper and much less specialised. These are the robot vacuum cleaners and swimming-pool cleaners, and the robots used in education or entertainment. There are around 6.7 million of these and that number is growing all the time.

Along with jetpacks, moon bases and food pills, we were promised robot servants in our future but the closest we've got to that is a small, disc-shaped unit that beaches itself on the edges of rugs. Now suddenly we're supposed to be worried about the introduction of sex robots? Hmmmm …

Two legs bad?

From February to September 2017, the Science Museum in London ran an exhibition charting the development of humanoid robots. Humanoid robots are those that have human-like features, though not necessarily realistically so. The more realistic versions are known as androids (if they appear male) and gynoids (if they appear female).

The *Robots* exhibition began with birth. On entering the gallery the first exhibit directly in the line of sight was an incredibly convincing animatronic baby, upright against a wall and enshrined in coloured light, slowly moving its limbs. The baby was originally a movie prop. It would probably be extremely convincing in context, in a filmed scene, but there, in the dim light, pinned to an otherwise empty wall like an insect on display, it was creepy and disturbing.

The narrative of *Robots* was split into five sections by date: *Marvel* (1570–1800), *Obey* (1800–1920), *Dream* (1920–2009), *Build* (1940–present) and *Imagine* (2000 and into the future). At each stage, the story was framed in terms of our understandings of ourselves. *Marvel* explored the sixteenth-century notions of a world understood to run like clockwork. In *Obey*, the idea of the human worker as automaton was mirrored in the mechanisation of factory work. *Dream* drew together science and science fiction, examining our hopes and fears around our mechanical counterparts. *Build* explored the cybernetic age and the work of the past 75 years that has brought us to this point. Lastly, *Imagine* examined the human–robot interaction we might expect in the future and speculated on what our relationships with these machines might be.

The highlight of *Robots* was the way in which it brought together a unique collection of over 100 humanoid robots under one roof – one of the best collections on the topic to date. The chance to wander through aisles of androids and gynoids that I'd only ever read about was marvellous (and the chance to go wild in the gift shop afterwards to buy all manner of robot-themed souvenirs was also much appreciated). The timeline shown in the exhibition beautifully illustrated just how far we've come with making our stand-ins.

* * *

The first full-scale, digitally controlled bipedal anthropomorphic robot to be developed was the WABOT-1, released in 1970 by Waseda University in Japan. It looked as if it had been made from a giant Meccano set, all blocky metal panels and rivets. It did have a semblance of human form though: boxy, angular arms and legs with artificial eyes, ears and a mouth.

Honda pushed things forward in 1997 with their bipedal humanoid robot P2. Not only was its movement realistic but it also seemed more complete than its predecessor, having a more solid form. This subsequently led to their development of ASIMO in 2000. ASIMO (Advanced Step in Innovative MObility – and a nod to a very relevant sci-fi author) could interact with humans through recognition of objects, gestures, faces, sounds and body language. Laser sensors detected the ground surface and enabled ASIMO to walk navigable paths. It could even manage stairs.

Since ASIMO there have been a number of humanoid robots developed both commercially and for research purposes. Some, like Hiroshi Ishiguro's Geminoid series, are intended to be ultra-realistic; others, like Aldebaran's NAO, are small, open-source and cheap enough that they can be used widely for research and educational purposes. Robots like the iCub – a humanoid robot the size of a three-year-old child – were designed to test hypotheses. The iCub was built to interact with its environment and learn about the world in the same way that a small child might.

But why build robots that have a human form? We have plenty of robots in use that aren't human-shaped. It's an incredibly difficult and costly task to get robots to do the physical things we take for granted: balance, walk, pick up things or touch with varying degrees of pressure. We know we can have social interactions with non-human robots so, other than our sci-fi visions, why invest so much time and money in such a difficult task?

One reason is that it's the way our world is set up. We have built environments suitable for humans: human-sized doors and paths, stairs the right size for human legs and shelves within the reach of our arms. It would be quite handy to have robots that could fit into our environment. 'Why the human form?' asks a character in Isaac Asimov's enduring sci-fi work *The Caves of Steel.* The answer given is that it is the most successful form in nature and that it's easier to design a robot to fit our world than to redesign all our tools for them.

In Asimov's 1942 short sci-fi story *Runaround*, he presented the Three Laws that the robots in his fiction had to obey. These are: 1) that a robot won't harm humans or allow one to be harmed by failing to act; 2) that a robot must obey humans, unless obeying them breaks the first law; and 3) that a robot should defend itself, unless defending itself conflicts with the first or second law. Asimov added a zeroth law at a later date, which stated that robots would not harm humanity, or allow it to come to harm by not acting.

In an interview with *Compute!* magazine in 1981, Asimov remarked that the Three Laws were 'the only way in which rational human beings can deal with robots – or with anything else'. Alas, his laws are subject to misinterpretation, something he knew perfectly well as he wrote a lot of short stories that used loopholes in the laws as a plot-line. The laws are a great starting point for ethical decision-making, but there's simply no way of replicating them in a machine in a foolproof manner. So far, the only thing we can be sure robots can do is obey orders. No set of ethical rules can cover all eventualities in the real world. How would we even begin to program them when there are so many ambiguities? We can't even agree on a universal ethical code for humans. The United Nations' Universal Declaration of Human Rights is probably the closest we've got, and it's a list of desired outcomes, not a path to achieving them.

Quite a few groups of people have had a go at coming up with a code of ethics along the lines of Asimov. They realise that even before trying to make our robots *behave* ethically, we should be trying to ensure the ethical *development* of robots. We need human rules, not robot ones. In September 2010, a group of experts met in the UK to draw up general principles.* They came up with rules for the robotics industry, not the robots themselves. These included: not designing robots for use as weapons, except in the interests of national security; making sure robots comply with existing human laws (such as privacy); meeting safety standards; and making sure someone – someone human – has legal responsibility for their robot. Interestingly, they also said that robots should 'not be designed in a deceptive way … their machine nature should be transparent'. Humanoid robots are clearly

* www.epsrc.ac.uk/research/ourportfolio/themes/engineering/activities/principlesofrobotics/

machines that borrow some basic elements of the human form, but what about android and gynoid robots? Is it deceptive to try and build a realistically human-like robot?

In March 2016 in Austin, Texas, Hanson Robotics revealed their social gynoid, Sophia. Since then Sophia has rarely been out of the news. Sophia is said to be modelled on the actress Audrey Hepburn, although that's not quite evident on first glance. Sophia can produce various expressions and uses voice recognition and AI to carry out interactive conversations. Sophia was very deliberately designed as a robot with an acceptably human face: the back of the head is clear and internal wiring can be seen, but the face is intended to seem naturally human. David Hanson, the creator, has said that Sophia would make a suitable companion robot and could work in healthcare, customer service and education.

Sophia has been wheeled out – literally – as a guest on talk shows and news programmes and has even been introduced to the United Nations. Most controversially, in October 2017 Sophia was granted Saudi Arabian citizenship. This was the first time a robot had ever been given a nationality. It was not entirely well received, given that it came from a country with an appalling track record for human rights, where women face gender apartheid and where non-Muslims cannot be citizens. Hanson joked with US talk-show host Jimmy Fallon that the robot was 'basically alive' – a comment that led to uproar among those aware of Sophia's many limitations. The artificial intelligence ethics community has been scathing on the subject of Sophia. Terms like 'obviously bullshit' and 'Wizard of Oz' have been used. I agree with the criticism: Sophia may have a human face and upper body but is essentially a chatbot running off scripts.

'Most people don't care,' retorts Ben Goertzel, the Chief Scientist of Hanson Robotics. In his view, even when people know Sophia doesn't understand what's going on, they still want to relate to the robot as if it were a real character. Perhaps. But if so, are Hanson Robotics being irresponsible in pushing the notion – or, at the very least, not disabusing people of the notion – that Sophia is a clever and lifelike machine? Leading roboticists like Joanna Bryson and Yann LeCun find it problematic. Bryson is concerned that Sophia is a 'massive deception' – a convincing puppet that exploits people's naivety around artificial intelligence and robotics. As one of the people who rewrote the Asimov rules in 2010, she feels it contravenes the guidelines around machine transparency.

While people might unwittingly assume Sophia has some sophisticated intelligence, there's no mistaking her for a real human. Sure, she looks good, for a gynoid. But we are a very long way from the convincingly human-like robot.

The depths of the uncanny valley

In 1970, robotics professor Masahiro Mori came up with the concept of the 'uncanny valley' (originally *Bukimi no Tani Genshō*). His hypothesis was that we humans empathise with machines that have human attributes, up until the point where they approach indistinguishability from reality. At that point, the 'almost-human' robot becomes unsettling and uncanny, and fills us with revulsion. The origins of the term itself go back to early twentieth-century psychoanalysis. In 1906, Ernst Jentsch's essay *On the Psychology of the Uncanny* mentioned a particular instance of an unsettling feeling: 'one of the most reliable artistic devices for producing uncanny effects easily is to leave the reader in uncertainty as to whether he has a human person or rather an automaton before him'. Freud expanded on Jentsch's work in his 1919 essay *The Uncanny*, a discussion around the German word *unheimlich,* connoting the unfamiliar and the unknown. Freud being Freud, his discussion eventually wanders into Oedipus, the castration complex and fear of childhood, but not before a very thorough exploration of the cross-cultural language terms that convey this feeling ('*lugubre*' is my favourite).

The uncanny valley hypothesis is somewhat controversial. It's a subjective phenomenon, after all. People feel the sense of unease to varying degrees, and a 2015 academic paper by Jari Kätsyri and colleagues found that empirical evidence for the effect was ambiguous, if not absent. It may well be a culturally influenced phenomenon. Most people I've spoken to, though, have recognised some sort of strange unsettling prickle at one time or another. You might have watched a horror film that has played on this: masks and mannequins inducing terror in the alive-but-not-alive form.

There are a number of theories as to why we experience this feeling of the uncanny valley. A significant factor may be that 'human-looking but not alive' is redolent of death. Mori gave the example of viewing a corpse: it's not a pleasant experience and could be frightening, and if it were to move then we would be terrified. For robots, it isn't just the realistic-but-not-quite-human appearance that can disturb us. Coupled with not-quite-human movement, a new dimension of uncanniness sets in. Could the uncanny valley be an unwanted reminder of our own mortality? Are we humans hardwired to reject signs of death?

Conflicting perceptual cues cause cognitive discomfort. The current humanoid robots often have mismatched actions, such as speech that doesn't match mouth movement, or facial expressions that don't convey a mood that matches the conversational tone. We are still stranded in the uncanny valley. Current human-like robots could never be mistaken for a real person. It may be that they never will be: the 'uncanny wall' theory proposed by researcher Angela Tinwell suggests that, as technology develops, we will become more and more attuned to the subtle mismatches between human and virtual. We might never bridge the divide. It may be easier, some day, as technology advances. For now, it remains a very difficult task. Human-like robots may dominate the media but they are still a long way from dominating the world.

Care and companionship

Two thousand years after Laodamia's tragic proto-sex-robot story was written, a French robotics company and a Japanese telecommunications corporation teamed up to build a new humanoid prototype – a companion robot named Pepper. The collaboration by Aldebaran Robotics and SoftBank Mobile was a venture to create a robot that could respond emotionally to the people it encountered. Pepper was designed to analyse human voices, facial expressions, body movements and words, and to react in a natural and appropriate manner. This, claim the companies involved, means Pepper is a 'genuine day-to-day companion robot'. A more cynical observer might note that Pepper's deployment in retail stores and in interactions with customers means that a huge amount of data about consumer behaviour can be tracked and scrutinised. Nonetheless, Pepper is in high demand by commerce and the public alike.

Pepper is around 120cm (4ft) tall, made of shiny white plastic, the torso of a human with a solid, curving pillared lower half that moves around the room seamlessly on a wheeled base. The eyes are wide and large and blink at you – a design feature that is known to engage us, psychologically attuned as we are to respond to the babylike traits that we describe as 'cuteness'. Pepper's voice, too, is childlike. We are meant to infer safety and trustworthiness. The eyes and mouth, however, house the cameras that allow Pepper to gather the information needed to analyse our emotions. Make no mistake: behind that charming exterior lies an emotionless machine whose entire purpose is to help you express your feelings. The thing is, it works. People like Pepper. People want to interact with Pepper. Just a few years after its launch you can now buy your own Pepper to become your home companion, integrating into your life as a friendly figure who is always there for you, who will always listen and who will always respond.

I've met instances of Pepper on several occasions, usually in museums or at trade shows, and recently at the Norwegian Developers Conference, NDC Oslo. Pepper was there on the exhibition floor, not advertising companion robots, but as a way of attracting interested and potential customers to an unrelated software product. Pepper offered to dance for me, so I agreed. Music began, gentle and swirling, and Pepper's arms gracefully extended, flexed and turned, its head bowing and lifting as it followed the movement of the hands, the body turning on its wheels. It did indeed look like dancing. The current commercial face of emotional companion robots is a humanoid machine of loving grace.

If you decide to buy a Pepper for yourself you should be aware that the terms and conditions have an interesting clause. You cannot use Pepper for 'Acts for the purpose of sexual or indecent behavior, or for the purpose of associating with unacquainted persons of the opposite sex'. The wording of this is vague – possibly intentionally so – and doesn't quite specify the, er, nuts and bolts of the specific activities. What does seem clear, however, is that the manufacturers are quite aware that someone, somewhere, might want to have sex with their robot.

In 2017, Pepper was brought into a nursing home in Southend in the UK, to interact with and monitor the health of the residents. Reactions were mixed. Critics immediately pointed out that there is no substitute for human touch and interaction. While they acknowledged the carer shortage in the developed world, they viewed the robots as a temporary solution, masking fundamental problems in supporting an ageing population. The elderly users, however, were more positive. The *Daily Mail* reported one care-home resident as saying: 'the most important thing is health ... My family don't live [nearby] so anything that can help me and my disabled husband is fantastic.'

Care robots have been around for some time. Japan has been developing robotic devices that can aid frail people or those with disabilities to move around their homes. Japan's robot strategy, published in 2015, talks of the routine use of robots in surgery and nursing to provide advanced medical care. Elderly care is an increasingly urgent challenge. Their nation's baby-boomer generation is reaching the age where care becomes more necessary: they have a top-heavy population that will require an estimated 2.5 million care workers by 2025. There is a labour shortage but Japan is reluctant to employ immigrant nurses. In 2015, Prime Minister Shinzo Abe spoke of the need to raise the country's birth rate. In the meantime, there is a push for technology to fill that gap.

Japan's robot strategy is not to introduce humanoid robotic equivalents of healthcare assistants. Instead, the idea is to develop better robotic nursing equipment such as robotic wheelchairs that can work independently and safely. The premise is to help those in need of assistance to lead self-sustaining lives. Okay, so there are *some* slightly weird robots. Robear is an experimental care robot: 140kg (over 300lb) of strong, white, bear-shaped plastic just slightly smaller than a human. Robear's purpose is to lift and carry patients. This could make a dramatic difference to care-giving staff who need to lift patients on average 40 times every day, and for whom back pain is a very real problem. Robear is still very much in the trial stages. Despite its manufacturers, RIKEN, describing it as 'the strong robot with the gentle touch', it turns out that the delicate movements required are quite beyond its capabilities for now. It can't yet be let loose on humans, particularly fragile ones.

My grandmother, with whom I share a name, was a strong and independent woman who I adored. My favourite picture of her is one where she is in her twenties, sitting beside her sister in the yard of the house where she grew up, head thrown back, laughing. My childhood is filled with happy memories. She was a hardworking working-class woman not given to emotional outbursts, yet she always managed to show her love for her grandchildren. She was in her late eighties when she developed vascular dementia. It's the second most frequent type of dementia. Unlike Alzheimer's, which is the most common, vascular dementia is caused by interruptions in the blood flow to the brain. Owing to its nature, it often comes with problems such as impaired mobility or trouble with speech – similar effects to the damage caused by strokes, which can be a contributing factor.

My grandparents had bought their first home when I was 10 years old. My granny was incredibly proud of it. It was her territory: she was no longer paying rent to live in someone else's property. She had the security she had long sought. After the death of my grandfather, as her health declined, it became much, much more difficult for her to move around her home, to cook and to support herself. But she refused to leave. She was determined to stay in her own space. My mother, who lived 25 miles away, took on as much as she was able (she was also looking after my father, who had had a second stroke). She arranged carers and she employed someone to visit every day and make sure my granny was eating.

On her 90th birthday, I flew back to my home country of Northern Ireland to visit her. We talked, although she had trouble getting the words out: vascular dementia causes aphasia. But she made it very clear: she felt her life was at an end and, despite my mother's attempts to persuade her into a nursing facility, she would live it out in her home. She was 91 when she died. She had fallen and had lain there unable to get up and get help. She passed away while I was on the way to the airport, hoping against hope to see her one last time.

If I could have given my granny the ability to live in her own home, safely and independently, I would have jumped at the chance. She knew what she wanted and she was determined (stubbornness is a trait I inherited, pleasingly). It was of vital importance to her to spend her last years surrounded by the sum of her life. If robots could have made that easier for her, I would have welcomed it in a heartbeat. I think she would have, too. Robotic carers may sound like a cold, emotionless, somewhat heartless solution to a care crisis, but they could bring quality of life to people desperately yearning for independence. My granny could have benefited from robotic assistance in standing, walking, cleaning and cooking. Companionship would often have been welcome, too: she delighted in visits from her great-grandchildren (in small doses), and she had a cat, though she found looking after a pet difficult.

Companion robots, rather than robots purely for practical care purposes, are another possible solution to the problem of an ageing and lonely population. Naturally, there are compelling arguments that state that machines, no matter how interactive, could never – *should* never – replace human contact. There's a lot to be said for that. For many people, human contact is the ideal. In fact, the UK government recently appointed a minister for loneliness. Their commission into the extent of the problem found that 'more than nine million people always or often feel lonely', that 'around 200,000 older people have not had a conversation with a friend or relative in more than a month' and 'up to 85% of young disabled adults – 18–34-year-olds – feel lonely'.

There is no denying that our communities have changed as the world has advanced. We are a global society these days: families and friends are disparate, and people move far from the place they were born for study or work. The technology that has changed the landscape of employment has also allowed us to stay in touch with our loved ones, even across long distances. But is that enough? What if we don't have anyone to reach out to?

Recently there have been trials using robotic companions therapeutically. This area of development has emerged from a branch of science known as social robotics. For the past 25 years researchers have been developing robots that can interact on a social level. Pepper is just one in a long line of these. The first was Kismet, a robot head made by roboticist Dr Cynthia Breazeal in the late 1990s in Massachusetts Institute of Technology's AI department. Breazeal built Kismet as part of her doctoral work. It has some human features: false eyelashes and red lips on a mechanical framework face. Kismet was designed to learn about the world much like an infant would, via socially situated learning. The more interactions it has, the more information it learns, so that it can respond better and better to social-emotional interactions.

Through her work with Kismet and its successors, Breazeal learned not just about robotic development but also about people's interaction. She found that when people engaged with robots they formed an emotional relationship, much like the relationship with a pet. It's no wonder, therefore, that robot pets have become one way of providing companionship. Probably the most famous of these is Paro.

Only the hard of heart could meet Paro and remain unmoved. Paro is a fluffy baby harp seal with big brown eyes and little squeaks of excitement. The cuteness level is off the chart: Paro looks just like a big cuddly toy. Well, that's what it is. But as well as that it's a therapeutic robot. Beneath its fur are processors, microphones and sensors. Stroke it and it will close its eyes and twitch its body and chirrup at you. It can learn names and remember faces. It even has FDA medical approval.

Studies with care-home residents have indicated that interactions with Paro increases group participation, have a calming effect and reduce users' anxiety. In fact, in one study, residents interacted more with Paro – and with each other when Paro was present – than they did with the care home's dog. To add to the dog's woes, Paro doesn't need to be fed or walked, and no one has to clean up after it.

Despite the positive results there have been criticisms of Paro's use. Concerns have been expressed about deception, particularly among users with dementia, and negative impacts on their dignity. Is Paro tricking them? Will they look foolish? Or is it enough that they engage and we see benefits? These are questions that are firmly embedded in ethical discussions that still run on today.

Sherry Turkle has been studying human–technology interactions throughout her career. Her 1984 book *The Second Self* became a widely read and well-received classic in the field, analysing how people from all walks of life experience and engage with the technology they use. It's a fascinating read. Through case studies, she reports on people's willingness to form social and emotional relationships with robots, treating them as they would a pet or another human, even when they are aware it is nothing more than a machine.

This is a phenomenon we see time and time again. We relate to the world around us in terms of our own social framework. It makes sense to us to interact with responsive things as if they are capable of understanding us, even when we know they aren't. We respond to each other in that way; we respond to animals in that way. If a computer talks to us or sends us some sign that it is responsive, then our default way of engaging is one that is fundamentally human, with human expectations.

From sharpened stones to Silicon Valley, 3.3 million years of technological development has brought us to a world where we can offload labour onto machines. Along the way we've learned how to control fire, to put wheels on things, to store food, to weave fabric, to write, to capture pictures, to broadcast, to communicate across the world, to save lives and to take them. We have refined each of these achievements to the point where we can separate ourselves from the tools that carry out these actions. Humans have always used technology to increase physical productivity. Robots are just the next stage in that. And now, as we shall see, we're trying to offload the mental aspects as well.

CHAPTER THREE

On Paperclips, Cats and Zombies

Imagine a world where the biggest, most terrifying threat to humanity is the humble paperclip. Such an innocuous object, lying loose and buried in a drawer, or hiding amid the elastic bands and staples in a desk tidy, seems utterly benign and in no way menacing. That's exactly why artificial intelligence expert Nick Bostrom chose it as an example of how our attempts at making machines think could go terribly, terribly wrong.

Bostrom hypothesised the paperclip maximiser – an artificial general intelligence (AGI) whose entire purpose is, as the name suggests, to maximise the number of paperclips it has. Designed by humans as an assistive tool, its objective is to do this by any means: gathering paperclips, buying paperclips, stealing paperclips or making paperclips. Let's say it begins with a level of intelligence equivalent to human intelligence. It has one single focus – more paperclips – and so it expands its intelligence purely to fulfil its goal. The paperclip frenzy increases; the machine is unwavering. Nothing can get in its way, because to get in its way would mean preventing it from reaching its goal. Humans can't intervene to stop it: they are a threat too. Goodbye, humans! Soon all matter in the universe is turned into a paperclip.

The paperclip maximiser is thankfully just a cautionary thought experiment, and anything could take the place of the paperclip, but the message is powerful: when we create machines that learn and act, we need to be careful about what we tell them to do. Artificial intelligence (AI) is the concept of machines being able to carry out tasks in an intelligent manner. A key objective here is reaching the goal, not necessarily how that goal is reached, although varying branches of the discipline deal with the ways this might be achieved: for example, by mimicking human cognition, or by taking a purely computational, mathematical approach. At present, none of the AI in use today involves machines actually being sentient or conscious, nor do they have any general intelligence. Rather, they can use data to understand patterns of outcomes and in that way learn from previous situations.

These days, we're perfectly used to computers carrying out boring and laborious everyday tasks for us. This morning I awoke when my radio alarm gradually lit the room (and then unceremoniously launched into a news programme so I could become jaded and angry within five minutes – this stops me falling back to sleep). I checked my phone. I checked social media. I dragged myself out of bed and checked my child. In the kitchen, I microwaved a bowl of porridge for her. Off we walked to school, past banks of borrowable bikes, waiting to be released from their docking stations by entering a computer-generated code. Once my daughter was safely installed in her classroom (with its interactive whiteboard) I walked the few hundred yards to the station where I checked the electronic signs to see when the next train was due. One hour of daily life from the moment I opened my eyes, and computers of one sort or another were a seamless and ubiquitous part of it.

A brief history of thinking machines

You've guessed it: the Greek gods got there first. They started by creating Pandora, the first human woman. Her name, which translates as all-gifted (and all-giving), represents her creation by a mythical team of what were essentially godly programmers, each of whom contributed an attribute. Pandora is designed; she is an artificially intelligent agent.

The Greek myths are peppered with manufactured creatures that go beyond mere automata. As well as his robots, the god Hephaestus dabbled in AI, building an army of mechanical maidens – *Kourai Khryseai* – who could think for themselves. 'There were golden handmaids also who worked for him, and were like real young women, with sense and reason, voice also and strength, and all the learning of the immortals,' wrote Homer in the *Iliad.* They are unusual, reports Genevieve Liveley, in that 'they can speak, they have power of words and a kind of intelligence – that's a crucial thing. They go beyond being mere automata. I always imagined them,' she continues, 'looking like the robot from *Metropolis*.' This is an idea I enjoy. A fifth-century BC piece of pottery in the British Museum shows the creation of the first mortal woman, Pandora, standing upright with her arms by her side, flanked by two men, in a depiction that looks just like a scene from that film.

Outside of myth, machine intelligence was much more difficult. The automata created over the centuries could act, but not think or give the illusion of thinking. Von Kempelen's chess-playing Turk was the closest to giving the impression of cogitation, but the task of making a machine that could both think and move was far beyond the abilities of the time.

The English mathematician and engineer Charles Babbage saw early on the opportunity to mechanise calculations. In 1822 he began to design his Difference Engine – a machine that could automatically compute polynomial functions. It was never completed, but he was spurred on to design the next model, the Analytical Engine – a much more complex affair. It was a proposed mechanical calculator instructed via punched cards, based on the methods of the Jacquard weaving looms. Such looms took input in the form of these punched cards: if there was a hole in the card then the hook holding the thread was raised, and if there wasn't, the hook was stopped. Arrangements of holes and solid card meant different patterns could be woven with the rise and fall of the hook. Babbage saw the potential for this in computing, realising that mathematical instructions could be passed in the same way. Together with mathematician Ada, Countess of Lovelace, they wove algebraic patterns, writing algorithms to calculate sequences of numbers.

Alas, none of Babbage's engines were ever completed before his death in 1871, but he had laid the theoretical groundwork. His interest was not just in the process of automating mathematics. Babbage had also lost two games of chess to the Mechanical Turk and it got him thinking. He knew there was human trickery behind it but couldn't quite work out how. But he began to ponder how a machine might be able to think for itself, to compute chess moves. From working on his engines, he concluded that such games of intellect could be automated, and he began to devise strategies for doing so. It was to be more than a century before the next leap forward.

In 1950, the mathematician Alan Turing published his paper 'Computing Machinery and Intelligence', a landmark event in the science of the thinking machine. The field of computer science itself was still in its infancy, having been galvanised into action with codebreaking applications during the Second World War. Turing had described the principle of the modern computer in 1936 but it wasn't until 1944 at Bletchley Park, where Turing worked, that Tommy Flowers designed and built Colossus – the world's first electronic digital programmable computer.

Today, Turing has the accolade of being a founding father of Artificial Intelligence. Bletchley Park was the Allied codebreaking centre where mathematicians worked round the clock to decipher enemy communications. Turing's involvement was key in breaking encrypted messages, thereby providing vital intelligence that helped to shorten the war. In 1948 he joined the University of Manchester to work on what became known as the Manchester Computers, which included the world's first stored-program computer. From his early teenage years, following the death of his close schoolfriend, Turing had been interested in the nature of cognition, wondering how mind was associated with matter.

Turing began his paper with the question 'Can machines think?', immediately explaining within the opening paragraph that to answer that, one must first decide what is meant by 'machine' and 'think'. Not the easiest questions in the world – definitely longer-lasting than a meandering discussion over the third beer of the evening. And so he rephrased it, asking the reader to imagine a popular parlour game – the imitation game – where a person in a different room must decide, via written responses to the questions posed, if the answers they receive are coming from a man (A) or from a woman (B). Turing asked: 'What will happen when a machine takes the part of A in this game? Will the interrogator decide wrongly as often when the game is played like this as he does when the game is played between a man and a woman? These questions replace our original, "Can machines think?"' This simple question is the premise of the experiment we now know of as the Turing Test – our current universal test of machine intelligence. In the Turing Test as we know it today, a human judge must decide whether the answers to the questions they ask come from a human or a bot. To pass, the bot must persuade the average judge into making an incorrect decision at least 30 per cent of the time.

Turing was by no means the first person to ask this fundamental question. The French philosopher Descartes was particularly interested in the connection between body and mind and it became his key theory, known as *dualism*: the idea of a non-physical conscious mind, separate from the brain, linked but distinct. It was Descartes who coined the phrase 'I think, therefore I am' (in French, '*Je pense, donc je suis*', and in Latin, '*cogito, ergo sum*'). The phrase captures a fundamental argument about consciousness: if we are able to contemplate our existence, then we must exist, because how else could we do the contemplating? But Descartes did not believe that automata could think like humans. He envisaged a machine that could respond to speech or touch but that could never 'reply appropriately to everything that may be said in its presence'.

In the decade prior to Turing's 1950 paper, the idea of machine intelligence had been a topic of discussion among researchers. In fact, Turing himself wrote a report in 1941 exploring the idea of a machine that could show intelligent behaviour by playing chess. Turing's 'Computing Machinery and Intelligence' paper is so much more than the description of what was to become the Turing Test, although it was the test that captured the imagination of many. He also offered a critique of its value. He first explains the possibility of a digital computer capable of playing a part in his test, explaining that while it might not currently exist, such a machine is feasible and imaginable, and could be possible by the end of the twentieth century. He didn't specify what kind of feats this computer might achieve, just that it could respond as if showing understanding of the user.

Turing used the rest of his paper to discuss and refute nine arguments that contradict his assertions. The first is theological: God gave a soul to humans but not to animals or inanimate objects. He is placatory, saying that humans creating machine intelligence would be much like humans procreating, and therefore they would be instruments of God's will. The second argument, what he terms 'Heads in the Sand', asserts that the consequences of machines that could think would be 'too dreadful'. He dismisses this out of hand. Third, he details the mathematical objection that a machine could face questions it cannot answer. But, said Turing, humans are fallible too. We might be cleverer than a given machine but other machines might be cleverer than us.

The 'Argument from Consciousness', Turing's fourth argument, is where he slides into philosophy. It debates the test from a consciousness perspective: is the machine really understanding what it says, or is it just returning signals processed in some way? This is known as the 'problem of other minds', which I'll come back to later. 'I do not wish to give the impression that I think there is no mystery about consciousness,' remarks Turing, which is good because his test doesn't test for that. But, he says, it doesn't need to. His case is that if we can't tell whether or not anyone other than ourselves experiences emotion, then we should just go ahead and accept his test.

Argument five takes the stance that 'you will never be able to make [a machine] do X'. I am particularly fond of Turing's selection for X here. His examples are: 'be kind, resourceful, beautiful, friendly, have initiative, have a sense of humour, tell right from wrong, make mistakes, fall in love, enjoy strawberries and cream, make someone fall in love with it, learn from experience'. He explains that these criticisms are similar to the Argument from Consciousness. 'Possibly a machine might be made to enjoy this delicious dish, but any attempt to make one do so would be idiotic,' he writes about the strawberries. He doesn't elaborate at all on the whole falling in love thing, which disappoints me. I wish Turing had lived to see society's relationships with computers.

The sixth argument is that Ada, Countess of Lovelace, had said that machines could never do anything original, merely what it is commanded of them. No, says Turing. Machines can be surprising, even if it's just from working out unforeseen consequences from data – something that we shall shortly see is possible via a technique known as machine learning. The seventh argument is that a machine cannot mimic our complex nervous system. True, says Turing, but it doesn't need to for the sake of the imitation game – it can be mimicked well enough to pass. Once again this highlights that the game centres on the *appearance* of thinking rather than actual measurable cognition.

Argument eight is that human behaviour is not determined by a set of rules. If it was, then we'd just be a form of machine. In order to create a machine, you need a set of rules, so a machine can't be human. Turing argues that humans *are* governed by rules – the laws of nature. Just because we can't currently determine which laws or rules govern human existence does not necessarily mean that such rules do not exist, says Turing.

The ninth and final objection considered by Turing is that of extra-sensory perception (ESP). Yes, you read that right. Mind-reading. Sixth sense. At the time he wrote his paper, ESP was an area that had been the focus of scientific study, albeit study that would later be shown to be flawed, with results that were impossible to replicate. Ever the scientist, given there was (supposed) statistical evidence in its favour, Turing had to include the possibility of ESP impinging on the imitation game, though he does remark that he would like to discredit it. He argued, however, that 'a telepathy-proof room* would satisfy all requirements'.

Turing lays out how we might program a machine to win the imitation game by thinking of the development of a child's mind: nature and nurture, comprising hereditary material and education and experience with a learning process governed by reward and punishment. We could make a child machine, he says, and teach it, reassuring us that 'we need not be too concerned about the legs, eyes, etc.' His descriptions are compelling. 'An important feature of a learning machine is that its teacher will often be very largely ignorant of quite what is going on inside, although

* Or, as we know it today, 'a room'.

he may still be able to some extent to predict his pupil's behaviour,' he writes – something that holds true for systems today. But where to start with this? How to get from the first steps of reasoning to the simulacrum of humans? Chess, suggests Turing. And so it came to pass.

The tragic end to Turing's life at the age of 41 meant that he never saw his ground-breaking work reach fruition, but his paper became the foundation stone of thinking about Artificial Intelligence. His test may have its flaws but it enabled researchers to set a goal for AI. There was a target to aim for – one that, at the time of writing, seemed truly audacious. It still is. No machine has passed it yet. Over the years, however, the goal has shifted. Even if a machine were to pass the test, it might not tell us anything about that machine's intelligence. The abilities of a computer might not equate to the abilities of a mind. Has it really learnt anything? It may just be a machine that can perform well in a test that doesn't quite measure what we want it to.

Teaching machines to learn

Today, the AI research community is vast and global and not just limited to thought experiments. Getting tagged automatically in photos online? Yep, AI. Your supermarket offering you particular, personalised products? AI determined what they'd be. To do this, it needs data. Lots and lots of data. Fortunately (from an AI perspective), we generate massive amounts of data every single day.*

* If you'd like to find out more about this, I recommend Timandra Harkness's book, *Big Data*, which is also in the Bloomsbury Sigma series and also contains a high level of puns.

The most common way of implementing AI today is via machine learning. To explain this, let's consider it in one of the most popular Internet memes available: cats.

We humans are very adept at recognising a cat when we see one. Our exposure to cats has most likely occurred early in life – in picture books, photographs and simply by walking down the street. As we grow we learn to distinguish cats from other animals, such as dogs. Or rabbits. Or larger felines like lions and tigers. Humans can look at a scene and easily pick out discrete objects. We can tell if something is moving in the stationary landscape. We can identify multiple moving objects in a street scene, for example: cars, pedestrians, cats. If a cat is sitting on a chair we can tell which bits are cat and which bits are chair. Easy, right? So simple a child can do it. But a computer finds this much, much harder.

For years, computers had to analyse pictures by segmenting the image into parts. After all, the photographs and video of real life that a computer works with are two-dimensional. To a computer, a photograph is merely a flat grid of pixels, and a video is merely a sequence of photographs. And so, computer vision algorithms were built to do things like detect edges and analyse motion – to break the scene down into salient parts. Three-dimensional information was extracted from cues such as shading, contours and focus. Features could then be labelled by hand: they could tell the software that an object in a picture is a cat. The computer still has no idea of what a cat *is* but it's a step closer to knowing one when it sees it.

So we can program computers to recognise cats. Or dogs. Or items on a production line, or military aircraft in our airspace, or tumours in scans. It took advances in AI, however, to massively improve the success rate of this identification, and it did so using a technique known as machine learning.

Machine learning is a tool whereby we can give computers large amounts of data – known as datasets – and let the system learn to do things by itself. With machine learning, we don't necessarily have to give explicit instructions: the computer learns from the data, analysing patterns and looking for connections. Machine learning is a subset of AI. If we think of AI as the goal of making machines smart, then machine learning is one of the tools we can use to do this.

Those advertisements on social media that seem to read your thoughts? Those are the results of machine-learning algorithms trawling vast sets of data to tailor sales to your online history or your status updates. One thing you may notice, though: buy something online – a lawnmower, for example, and suddenly your screen is filled with adverts for variations on that item. It's almost as if the Internet wants you to start a collection of near-identical lawnmowers. Machine learning can be helpful but it's not necessarily *clever* – not in the way we hope our machines will be. Not yet.

Back to the cats. Machine-learning tools offer several methods for this all-crucial feline identification. In the first – supervised learning – the programmer gives the computer labelled data to analyse and shows it the correct answers. If we give a system plenty of correctly labelled images containing cats and images that are labelled 'not-cat' then the system can go through our images and spot the ones which meet our criteria for 'cat'. We have given the computer input in the form of a set of images and we have shown it the correct output (*i.e.* labels indicating which images have cats). The machine-learning algorithms have looked at the images and their labels, and tried to identify the patterns in the features that predict whether an image is indeed a cat. Once the algorithm has been trained on this correctly classified data it can go on to identify cats in new pictures. We are the teachers: we have supervised the learning process.

A word of caution, though. We have to be careful or our machine might learn the wrong thing. There's an apocryphal tale of a military tank-detection algorithm that was inadvertently taught to attack trees. The story goes that all the pictures of camouflaged tanks had been taken in a forest; non-camouflaged tanks had not. As a result, when an image had a 'yes, this is a tank' label, the algorithm also associated it with a forest and so solved the classification problem by looking for trees. The dataset wasn't an effective one. Unintended biases can be a big problem for machine-learning algorithms. The computer will pick up all our habits, good and bad.

An alternative method is unsupervised learning: the computer learns on its own. Once again the programmer loads the system with data, but this time the data is unlabelled. This time, the algorithms look for patterns and categories. The system has not been told what is a correct answer. Instead, it has to base its judgement on similarities and correlations. It can look for clustering of features, for example. As it whirrs its way through image after image, it might spot that these pictures contain fluffy objects with legs and pointy ears. It is looking for interesting and meaningful distributions. This is a really useful technique when we don't know what patterns or groupings might exist in a dataset. If we were analysing supermarket loyalty cards, we might use unsupervised machine-learning algorithms to make associations in customer purchasing behaviour, and the system might identify patterns we'd never spotted before because we didn't know to look for them. It's unsupervised because we've just let the machine get on with the work itself.

There is a halfway house between these two methods known as semi-supervised machine learning. As you might expect, only some of the answers are provided. We label only some of the dataset. So, for example, a photo album might have a few pictures of cats labelled as such but the majority of photographs have no label at all. We are teaching the machine by working through some examples but then we're giving it homework to do on its own. There are some parallels here with human cognition: parents point out objects to their babies and the babies work out associations and classifications through their everyday observations.

Another type of machine learning – reinforcement learning – is based on interaction and rewards. The algorithms used in reinforcement learning enable a system to identify suitable actions in a given situation in order to reach a goal. The system relies on a reward signal. As I write this, I'm engaging in a bit of it myself. If I hit this morning's 1,500-word target then I will reward myself with a cup of tea and (more importantly) a slice of cake. If I complete that action and am rewarded, my action was good. If I fail by deciding to take a nap, for example, then my action was bad. No cake. By associating a successful period of writing with the opportunity for cake, I am more likely to hit my word count.

Supervised learning is *instructive* (the computer is told how to decide on an output), but reinforcement learning is *evaluative* (the computer receives feedback on how well it has done). Your system could say 'I think this is a cat' and you could award a score out of 10 for how well it's done. This makes reinforcement learning ideal for games. It doesn't need rules: it just needs to play. The information required is a state (knowledge of the environment, such as where all the pieces are on a board), an action (where it might move next) and a reward (such as scoring points or making a winning move).

We can thank cats for reinforcement learning. Pioneering American psychologist Edward Lee Thorndike was the first psychologist to use non-human subjects. He wanted to know if animals (and, by extension, humans) could learn by observing which actions led to desired rewards. In 1911, about 25 years before Schrödinger hypothesised cats in boxes for other scientific reasons, Thorndike built puzzle boxes where an animal could escape by triggering a certain action and receive a reward. And so he put some cats in some not-very-big boxes. The cats, obviously, had no idea how to escape. They wandered around inside until they accidentally triggered the mechanism for release. But Thorndike discovered that in successive trials those cats got quicker and quicker each time at escaping, demonstrating a learning curve. Learning could be reinforced.

The scope for reinforcement learning is huge, leading to systems that can interact with their environment and make the best decisions in a situation. This has profound implications for progress in robotics or self-driving vehicles, for example. Imagine a robot that is aware of its position in a room and can make decisions about how to negotiate obstacles, and can reach out and grasp objects without being programmed to do so. Reinforcement learning could well be a part of artificial general intelligence, if that ever happens.

Our biological brains are capable of great things. We navigate our way through everyday life with them, interacting with the world around us and continuously solving problems, from where to place a foot on the ground in order to take our next step to tackling the thorny problem of how to get from A to B without a maps app on our phone. All the time, neurons – the nerve cells in our brain – are carrying electrical signals that tell our body how to respond. Neurons are the basic information-processing unit in the brain. Most animals have neurons, with the exception of sponges and some other simpler, multicellular organisms. A roundworm has 302 neurons. A cat has about 760 million. Humans have around 86 billion. Individually, they transmit nerve impulses. When they fire a signal, they connect to an adjacent neuron via a gap known as a synapse. Neurons have inputs – stimuli from our sensory organs – which transport information about our surrounding environment. These inputs pass through networks of neurons, processing information. The output is a response by our nervous system.

At a very, very basic level, we could think of neuronal signalling as gates that open and close: they fire and spike, or they don't. This seems handy, as circuits and digital computers also have their basis in two states: on and off. One and zero. Open and close. But that similarity doesn't hold out for long. Although the electronic signals in the brain could be said to use this basic language of all or nothing, the information processed can't be fully represented by computer bits. The spikes in neural code happen in rapid bursts, unlike the stable switching between on and off of a digital circuit. The systems in the brain are continuously working in parallel, without any central control. There are layers upon layers of fluctuating and adapting signals – millions of input events every second – and the rate at which neurons fire is of fundamental importance. A neural effect involves an ongoing chain of signals. We don't even fully understand the nervous system of the poor roundworms with their mere 302 neurons. There is simply no way we can currently mimic all the vast complexities of the human brain in a computer.

Creating an artificial neural network equivalent to a human brain is, therefore, more than just a little bit tricky, but it can be done simplistically with very useful results. If we consider a neural network from a mathematical perspective, then it is simply a list of operations to be carried out on a given input.*

A simple artificial neural network has three parts made up of processors: an input layer, a hidden layer (or layers – you can have lots of hidden layers if you want it to be less simple) and an output layer. Each input has a value. An output is calculated by passing the input values through a computation in the hidden layer, which tells the 'neurons' whether or not to fire. Let's think of the cats. Suppose you've been teaching your artificial neural network by showing it pictures of cats. Suppose there are five input nodes: one for head, one for body, one for legs, one for fur and one for pointy ears. Their presence is categorised as a binary yes or no. If all five are present then it's a cat (I realise this involves some very elementary assumptions; you can program your own, more realistic version later if you like). Now present your neural network with a new picture: a wooden chair, for example. Does the chair meet the input categories? Head? No. Body? No. Legs? Yes. Fur? No. Pointy ears? No. The hidden layer is looking for a threshold value of five because five means 'cat'. Your chair characteristics don't meet that threshold value. Hey neuron, it's not a cat: don't spike!

* Disclaimer: I use the words 'simplistically' and 'simply' on a comparative basis. The maths behind this stuff terrifies me.

You might think I'm just being whimsical with the cat examples but I'm not. Google's got my back on this. In 2012, their computer scientists hooked up their face-detecting algorithms and artificial neural networks to YouTube and set them to work, without supervision, to see what they could learn from 10 million cat videos. The results were ground-breaking. They published this as a bona fide academic paper, 'Building High-level Features Using Large Scale Unsupervised Learning'. They had shown for the first time that a computer could be given huge amounts of data and, from that data, could pick out a pattern we humans know as 'cat' without ever having been told what a cat looked like in the first place. It had developed its own internal representation of 'cat'.

Google's breakthrough came by scaling things up massively: much bigger datasets, more hidden layers, hugely increased processing power. The YouTube cat detector had 120,000 inputs, one each for three colour channels of the 200x200-pixel thumbnails. It had over a billion connections spread over nine layers of network. The huge size of the training set and the network were what helped it succeed. The depth of the layers make it an example of what is known as deep learning – a type of machine learning that uses multiple layers of processing, where each layer uses output from the previous layer as its input.

The concept of deep learning has been around for several decades – the phrase was coined in 1986 – but it wasn't until around 2010 that interest soared. Improvements in computer hardware meant that processing time could be reduced from weeks down to mere days. Multiple layers meant that machine learning could go deep. This depth meant better mappings between the inputs and the outputs of the neural networks. To produce generalisable and consistent results, much bigger datasets were required – such as Google's

10 million cats. Interest began to grow: the scope for applying this technology to applications such as speech recognition, image classification and language translation was apparent. There was clear commercial interest from corporations such as Facebook and Amazon. Let's skim over whether or not that's a good thing …

Deep reinforcement learning is the reason DeepMind's AlphaGo computer program beat a (human) professional champion of the fiendishly difficult Chinese game Go at least five years before anyone thought such an achievement would be possible. In October 2017, DeepMind released AlphaGo Zero, which took only three days of training to be able to beat the Go world champion. It did this by playing itself, learning from each game and improving with each iteration, growing the ability to predict opportunities and threats. Not content with that, they then took the same basic architecture and gave it the rules of chess. In just four hours AlphaGo Zero not only taught itself to play the game but surpassed all humans and every other chess computer program ever written. In a 100-game match it thrashed the reigning computer chess champion Stockfish by 97 to 3.

Deep its learning might be, but a computer judging our pictures still doesn't know what a cat is. It can spot one but it has no concept of its existence. It has no understanding of the animal 'cat'. It is smart enough to tell the difference between different animals, different road signs, different speech sounds and different languages but it doesn't *understand* these things – it has merely worked out the rules to classify them. It presumes a stable world with a certain degree of consistency. It's all just mathematics. It will never involuntarily laugh at a cat on a robot vacuum cleaner dressed as a shark. It'll try to sell you a robot vacuum cleaner over and over again even after you've bought one, though.

Voice boxes

Back in 1966, Joseph Weizenbaum, a German-American professor of computer science, developed a program called ELIZA that could recognise cue words and output corresponding conversation. The ELIZA chatbot gave the illusion of comprehension. Behind the scenes, ELIZA was responding to keywords as dictated by a script. The program checked what the user typed in and used pattern matching to return an answer from a list of pre-written and deliberately vague replies. In one case, Weizenbaum formulated a program that emulated psychotherapists' responses, replying to a user's statements with pre-programmed prompts such as 'Why do you say this?' and 'Go on, don't be afraid'.

Weizenbaum wasn't trying to fool anyone with ELIZA. It was, in his words, a parody. He wanted to show just how superficial human–machine interactions were. There was no machine learning involved; there was no language processing happening. It was merely template-based. But Weizenbaum was shocked by people's responses: many users assumed intelligence where there was none. People forgot they were talking to a computer. Without intending to, he had created the illusion of a humanlike machine.

We humans use language as our predominant method of communication. We don't know how language evolved. There aren't any traces in the archaeological record other than suggestions, such as the presence of a larynx. The larynx houses our vocal cords. When we speak we produce a sound wave. Our throats, noses, mouths and lungs shape that sound wave. It vibrates the eardrum of the listener, and the eardrum transmits those vibrations through the middle and inner ear and on to the brainstem. The brain then perceives this as a sound. The actual mechanisms of this are not fully understood, particularly when it comes to speech perception. So, even before we get to the level of hearing actual words, we don't quite know what's going on in terms of auditory processing.

If you've ever encountered the saying 'If a tree falls in a forest and no one is around to hear it, does it make a sound?', that's a philosophical musing on the nature of perception. A tree can fall over and propagate a sound wave, but if that sound wave is not interpreted and transmitted to a listener's brain, is it really a sound? The thought experiment goes much deeper than that, prompting discussion on reality and experience and the nature of existence, but let's just stick with sound for now, because no one wants to get into metaphysics without proper preparation.

Speech, therefore, is a series of sound waves that our brain can very cleverly interpret, filtering out background noise to comprehend not just words but context, tone and intention. It's stunningly complex. Our ability to make sense of language – to demonstrate logical reasoning and to express our inner emotions – has got us to the top of the food chain. We can take huge amounts of sensory information, process those signals, gain awareness of what we are hearing and react accordingly. By comparison, a computer can take the same signals but has no idea of the intrinsic meaning of them. Furthermore, signals aren't consistent either, so it's not simply a case of waiting for a universally recognisable one. If you ask people to say the same word, the speech signals they produce can be vastly different, especially taking into account things like background noise, emotional state, age, speed, volume and pronunciation. A machine needs to analyse and categorise those signals. It can recognise patterns and can search for context, but it isn't thinking. We humans can read between the lines. We can even infer meaning from things *unspoken*. A computer can't. It isn't making sense of words the way we make sense of words.

But even the limited abilities of a computer to process speech can be fantastically useful. Because it is the primary way for humans to understand each other, it makes for a very natural method of interaction. A screen-based interface requires a graphical representation of a command. Those aren't always easy to represent and they require more work from the user, who needs to seek out the appropriate icon or word in a menu and then check that it matches what they want to do. By contrast, issuing voice instructions doesn't require a graphical representation that someone else has designed. It lightens the cognitive load on the user. That makes it far more efficient – if it works correctly.

Fifty years on from ELIZA, millions of households give orders to machines just by talking to them. And those machines talk back. The popularity of virtual digital assistants such as Amazon's Alexa, Apple's Siri, Google Assistant and Microsoft's Cortana shows just how easy we find it to speak naturally to software. It's been forecast that by 2021 the number of AI digital assistants in the home will equal today's global population.

The difference between today's digital assistants and Weizenbaum's ELIZA is speech recognition and advances in natural language processing. ELIZA handled typed input, but today's natural language processing is able to deal with voice input. Attempts in the late 1960s and early 1970s to control computers using verbal commands were ambitious and were limited by the primitive hardware available at that time. A 1970s project by the US Department of Defense led to a system that could understand just over 1,000 words. This was a big jump from the decade before when only numbers spoken by a sole individual could be recognised.

In the 1980s, better hardware combined with the use of statistical probability pushed speech recognition forward.

This meant that context-dependent models could be used: the algorithms could calculate the likelihood of a unit of speech based on the speech it had already heard up to that point. Initial approaches in automatic speech recognition involved speech-to-text as a method: convert the waveform into words. But this proved problematic. To do this, the system needed an established vocabulary so it could identify the spoken words. The trouble was, potential users tended to speak in natural sentences, which often included words that weren't in the vocabulary. Moreover, the system was entirely one-way: the user spoke and the system listened. That goes against the grain of human-to-human communication, which is very much a two-way process, even if aspects of that process are non-verbal. We expect feedback in our interactions. Without feedback, we have no idea if the system is working. As a solution, a query–confirmation dialogue was introduced. Users could be prompted (for example, 'please say your account number after the tone') and an action could be confirmed (such as repeating the account number back to the user).

In 1987, Texas Instruments brought automatic voice recognition into the home in the form of World of Wonder's Julie doll – a battery-operated 'doll that understands you'. Julie had a cassette tape inside with recordings of responses that could be triggered by eight keywords: 'Julie', 'yes', 'no', 'okay', 'hungry', 'be quiet', 'pretend', 'melody'. (I always thought 'melody' was an odd choice. What child would choose to say 'melody' as opposed to 'song'?) The Julie doll also contained sensors that responded to temperature, light and movement. As she spoke, her eyes and mouth moved. Setting up the Julie doll involved a process similar to that for digital assistants today: she asked the user to repeat her command words in order to establish recognition of that user's voice.

By the 1990s, automatic speech recognition software became a commercial reality with the release of Dragon Systems' *Dragon Dictate* – software that still leads the way in the marketplace today. It was expensive, it required a set-up period to train it on the user's voice, and the user was required to enunciate clearly and make a definite pause between each word. Nonetheless, it went on to become the world's bestselling speech recognition software.

Meanwhile, the development of typed conversations with human-like chatbots continued, with developers competing to be winners in versions of the Turing Test. In 1990, American inventor Hugh Loebner launched a competition for the most human-seeming chat program. Twenty-eight years later, the Loebner Prize competition is still running, with inventors attempting to fool human judges that their chatbot is really a person. Judges converse for 25 minutes with two hidden entities via typed conversation on the screen. They then have to decide which is human and which is the AI. Monetary prizes and medals are awarded in accordance with the ranked scores. There is a prize of $25,000 for a chatbot that the judges cannot distinguish from a real human and which convinces the judges that the human is actually the computer. No one has ever won this. Nor has the grand prize of $100,000 ever been awarded. To win the big bucks, the chatbot has to be able to understand text, visual and auditory input. A winner is unlikely any time soon.

The Loebner Prize has been criticised as being nothing more than deception and trickery along the lines of ELIZA. It is very much a product of the imitation game rather than an attempt at an intelligent machine. It has been dismissed by researchers, including AI pioneer Marvin Minsky, as nothing more than a publicity stunt. The judges are not experts and the conversations are superficial. Nonetheless, the Loebner Prize is interesting in the way that ELIZA was interesting: it indicates that true intelligence is not necessary in order for humans to feel they can converse naturally with a machine. A semblance of human-like behaviour can be enough for us to assume a degree of sentience. Hold that thought; we'll return to it later.

* * *

The past few years have seen the rise of the virtual digital assistant. Apple's Siri was the first of the modern-day versions of these in 2011, introduced as an iPhone 4s feature. Siri works via natural language voice commands, responding with answers to questions or performing actions to launch apps or search online. Likewise Siri's compatriots, Amazon's Alexa, Google Assistant and Microsoft's Cortana. When you talk to them, the algorithms scan your request so that they can isolate the command. The vast amounts of data they collect through your interactions with them, and through the information they have on their servers, can be analysed to optimise their answers. Machine learning takes this massive dataset and uses it to generate models for understanding the natural language commands. There are limitations, of course. Virtual assistants can't answer just any request. They are still hugely constrained and nowhere near anything resembling general AI. They are better and faster than their predecessors at recognising what people say, but they aren't perfect.

I experimented with this chapter: I'm dictating it to the software on my laptop using the laptop's inbuilt accessibility function, which recognises the words I'm saying aloud and types them straight onto the screen in front of me. The problem I have encountered, though, is that it can't really handle my Northern Irish accent. In fact, in the previous sentence it heard 'aloud' as 'and lied'. This is a common failing of speech recognition systems, and it leads to an interesting phenomenon: we have to adapt our speech to the machine. It's not just accents, either. Even our speech syntax is changing. If we want to interact with a digital assistant we have to begin our sentence with its name. 'Hey Siri, play the *Hamilton* soundtrack.' We don't have to say please or thank you either, although surveys show that many people do.

The big corporations making these devices are keen to have them sounding as human-like as possible, but perfectly so. There is no room for anything more than the tiniest bit of hesitation, repetition or deviation. Your digital assistant can't be *too* casual. Humans have a tendency to mimic speech patterns in conversations. Today's conversational interfaces are pretty good at recognising commands in natural language sentences. There is really no need for us to couch our orders in polite questions. Brief imperatives would work just as well. Virtual assistants are essentially just voice-controlled search engines. Asking 'Cortana, what is the weather like in London today?' is merely a long-winded way of saying 'Cortana weather London'. So, do we want a tool? Or do we want a friend?

Virtual assistants all have one glaring thing in common: they were all originally designed to have a female voice. Siri shook that off when launched in the UK and is now available in female or male, and American, Australian or British. More on the gendering later; for now, let's contemplate why they have human voices at all. Why not use a machine-like voice? And while we're at it, let's look at their idiosyncrasies. The manufacturers of each of these devices have imbued them with just a tweak of quirkiness that for all intents and purposes gives them a personality. You can ask them to tell you a joke or sing you a song and they will oblige in their own particular way. You can flirt with them; they each handle that with slightly differing approaches.

This deliberate choice to make the assistants seem human is based on research from human–computer interaction by Clifford Nass, who studied how users form bonds with their computers. He observed that people treated computers with female voices differently from those with male voices. He also revealed how humans can be just as voice-activated as the conversational systems they use. When using voice technologies, humans respond the same way as they would respond to another human in a social situation. It's not just limited to voice, either. Even interacting via text input is steered by this. In fact, his 1994 paper 'Computers are Social Actors' reported experiments that showed that computers only need a tiny set of human-like characteristics for users to feel a bond with them, even though the user is perfectly aware that the computer is merely a non-sentient machine.

Natural language processing doesn't stop at speech recognition and generation. Another beneficial use is sentiment analysis, also known as opinion mining. This is a way of analysing the affective state of the user. 'Affective' is a word from psychology that refers to emotions or moods. Our affective state can be determined in a number of ways: the emotions we express verbally, our tone of voice, our body language and our physiological responses, for example. If a computer is able to recognise our emotional responses then it can react accordingly, perhaps by recommending things or providing useful information, or just by gathering data, such as customer feedback. This is incredibly useful for businesses trying to market their products. In fact, sentiment analysis has been applied to areas such as financial trading in order to find out if the general social mood has an impact on stock market prices, or to determine the tone of the political landscape.

Sentiment analysis is pretty easy if all you have to do is analyse five online reviews, for example. Give them to a human to read and they will spot the tone of words, sentences, paragraphs and the entire post fairly quickly. But scale that up to 500,000 reviews and it becomes an impossible task without some form of computational help. It seems like a natural solution to look towards natural language processing for this help.

Sentiment-analysis computer programs can be as straightforward as looking for mood-based keywords such as 'happy' or 'sad', or associating various words with a type of emotion. There are word sets for this, such as the Affective Norms for English Words, a collection of 1,034 nouns, verbs and adjectives. Gauging emotion by keyword is complicated, however. It's easy to pick out positivity in the sentence 'My meal was good.' It gets more complicated with the phrase 'My meal wasn't bad.' The affect in the second sentence is still positive but is expressed with negative phrasing. Similarly, 'I wanted a good meal but I didn't get one' is confusing if the keywords lack context. Imagine the difficulty when sarcasm is thrown into the mix.

More sophisticated knowledge-based methods involve examining the context of the expressed sentiment, searching for the relationships between words, which might indicate a particular affective state. This approach yields better results but is much, much more time-consuming to annotate. Added to that, humans also have difficulty in classifying sentiment given all the subtleties of expression, disagreeing with each other's classifications around 20 per cent of the time.

When supervised machine learning is used, training documents can be labelled, for example, as 'positive', 'negative' or 'neutral'. Again, this is time-consuming to set up: datasets need to be cleaned to remove any unnecessary data, and then corresponding labels have to be added to identify the moods that are associated with the words and sentences. The machine-learning algorithm can then analyse the probability of these words and sentences being positive, negative or neutral in different situations. Then, when new data is input, the system can decide on a classification based on what it has been taught.

Human communication is fascinatingly nuanced and we are highly attuned to spotting it. Deep learning might take machines one step closer to identifying how we feel. The massive datasets needed are already out there: Facebook, Google searches, Twitter, Instagram, Snapchat, Pornhub. The uses for opinion-mining technology are countless. Virtual assistants can't yet perform consistent sentiment analysis but that doesn't mean the corporations behind them aren't working towards that goal. Amazon and Google, who currently make the most popular virtual assistants, are pushing their technology hard, integrating it into all sorts of devices: headphones, cars and even cookers.

Sci-fi television comedy *Red Dwarf* forecast affective assistants in the form of Talkie Toaster, an annoying, chirpy AI bread toaster with a 'personality chip', supposedly manufactured to provide breakfast-table conversation. Talkie Toaster could sense its users' moods but the one onboard the Red Dwarf spaceship was defective, and spent its time insisting its owner ate toast. When ignored, Talkie Toaster would sob hysterically and rant at the spaceship crew. Back in the real world, smart toasters are now a thing. They don't have personalities yet, and to be honest they seem a bit pointless, but stick a smart plug on any appliance and tah-dah! You could control it with your virtual assistant.

Do we want that? How comfortable are we going to be with robots that can sense how we feel and react to our mood? The technology is getting there. It's really not that hard to envisage a robot that could seemingly act emotionally. Today, people are walking around with their moves tracked – voluntarily or not – on cameras. We upload information to our social media profiles. We wear health trackers that count our steps and our heartbeat. If deep learning can be used to take that data, sift it for patterns, analyse our emotions and place them in context, then we have computers that can read us accurately and respond manipulatively – benignly or otherwise. If a machine can do all that, how much does it matter about whether or not it's alive?

Philosophical meanderings

Artificial intelligence may sound like an engineering problem, but it's an engineering problem that asks some very big questions. In fact, there is no way of separating AI from the domain of philosophy. We have to ask fundamental questions about our own lives to work out exactly what intelligence means. Currently, a machine does not know it is a machine. We can tell it that it is a machine but it is not conscious of the fact. We can try to model human cognition but we don't quite know how that works either. We haven't yet cracked what makes us tick. In our attempt to create human-like intelligence we are learning a lot more about our own minds, too.

Let's jump in at the deep end: what is intelligence? The word 'artificial' has a fairly simple definition. It means something that has been made or produced rather than something which occurs naturally. But intelligence? The short definition in the *Oxford English Dictionary* is 'the ability to acquire and apply knowledge and skills'. Well, computers can do that, to an extent. Clearly there is more to it than that. The word itself comes from the Latin *intelligentia,* which translates as 'comprehension' or 'perception'. This, then, implies some form of understanding. From a psychological perspective, the concept is broader and involves other features such as abstracted thinking and making sense of things. The American Psychological Association points out that when two dozen prominent theorists were asked to define intelligence they gave two dozen different definitions. The bottom line is that human intelligence is a cognitive process by which we learn and reason.

In research, the term 'strong AI' refers to the presence of consciousness in machines. That is the next sticking point: what is consciousness? Unlike the Turing Test, there's no test for consciousness. Intelligence is not the same as being sentient. If someone asks you 'Are you clever?' you can demonstrate to them just how clever you are. Maybe you could translate a story, or play some chess, or solve a cryptic crossword (kudos to those of you who can solve cryptic crosswords; I remain baffled by these). But, as you now know, a computer could do those things too. And if someone asks if you are conscious – how do you prove that? We all have a concept of what consciousness is, but it's very hard to put that into words and it's even harder to devise a test for it.

Conscious experience is something familiar to us all. Probably. But then how do we know? If we stand together in the same room, how do I know that you see the same things as me? If I see an apple in the fruit bowl on the table next to us, and we both agree that it is red, how do I know that what I experience as red is the same as what you experience as red? What does this 'redness' of red mean? What type of property is 'redness'? We could measure the colour of the apple to find out its physical properties, including its colour based on measurable wavelengths of light, but we still might each perceive it differently. Just because we've agreed on a name for the colour we see doesn't mean we are seeing the same colour. The very fact that I am writing this and you are reading this and we are both thinking about this shows that we have introspection, and introspection means that we are self-aware. Doesn't it? Or maybe it is just me who is thinking about it. How do I know if you have a mind? I mean, you appear to be as human as I am, but that could just be an act.

If you are a philosopher you can skip this next bit. I have no formal training in philosophy but I've had to read a lot of it while working in this field. I admit it often ties my brain in knots. I'm not entirely sure I've got to grips with it all, but I'll have a go. It's one of those fascinating and twisting subjects: a walk through a forest with paths that lead off in all directions, crossing and looping back or running on into a tangle of trees. We, dear reader, are poised on the edge, keen to explore but somewhat worried that we might become lost. I'd ask a philosopher to guide us but they can't always agree on the right route. They are very good at arguing their case for the best one, though.

I don't want to make this forest seem too frightening but I warn you – there are zombies ahead. Here comes one now. It looks like you; it looks just like you. It talks and moves like you. It could be your twin. It could replace you and no one would know.

Don't worry. This zombie is just another thought experiment. It's a philosophical zombie so it's not trying to eat your brain. The philosophical zombie lacks subjective awareness. All its actions are governed by physiology and behavioural responses. If you scanned its brain and yours there would be no difference whatsoever, even though you can experience things and it cannot. It is a mechanical meat version of you. You are physically identical right down to the last neuron. It could go to work and do your job just as you do and no one would be able to tell the difference.

The philosophical zombie is a thought-construct of a replica human: one that can smile a smile, yet feel no happiness, or break its foot, yet experience no pain. Actually, right now, in real life, I have a fracture-shoe on my broken left foot and can assure you that I am not a zombie, although fortunately I do have access to prescription painkillers that seem to turn me into one at bedtime. If the philosophical zombie did break its foot, it would have behaved in the way I behaved: its identical brain would have sent signals to its limbs telling it to stop, to limp and to swear. It could even tell you that its foot hurts. It doesn't; but the zombie could tell you that it does. Maybe I *am* a zombie. I'm not. But then, that's what a zombie would say.

Now think of the zombie as a human-like robot that is indistinguishable from a real person in appearance and behaviour. It reacts the way we expect a human to react. It can process thoughts, it has memory, it holds conversations with us. But it has no awareness. It's just doing what it has been designed to do. But it is such an advanced simulation that you can't tell it from a human. There is literally no outward difference. Does it matter, therefore, that it isn't conscious? Is subjective experience necessary?

If you think such a thing is different to a human then you must think that consciousness is a necessary component of being a human. The ideas behind this line of questioning go back to Descartes, but the zombie idea was reactivated by philosopher David Chalmers. He wanted to know why we feel something as a result of all the brain processes going on in our heads. For some reason, our brains enable the self-awareness that lets us imagine what it is like to be us.

Why do we experience things? How do the mechanisms of our bodies and brains give rise to conscious sensations? Where does that consciousness come from?

You don't have to have an answer in the Great Zombie Debate. The philosophers can't agree on it either, which is why you have a bunch of very clever middle-aged white men amusingly inventing words like 'zoombie' and 'zimboe' to put forward their own variations on the theory. For now, just hold on to it as a way of thinking about what makes us *us*. Can we create a machine that can think? Could a computer ever be conscious? The jury is out. Some researchers, vastly aware of human limitations in understanding our own cognition and perception, insist that it will never happen. If we don't even know what sort of thing it is in humans, how could we possibly know what sort of thing it would be in computers, and how could we hope to recognise it if it arose? Others feel that it's possible, but it may not resemble our own recognisable form of consciousness.

Research in human–computer interaction shows that users often treat computers as having a social role. Studies have shown that even in something as non-human as a word-processing package, users respond to positive praise from the software and are irked if the software reprimands them for making errors. I've done it myself. In fact, who hasn't sworn viciously at a crashing app or hurled insults at an unresponsive printer? And why do printers always seem to break down when they're urgently needed? I sometimes even get the sneaking suspicion that they are doing it deliberately.

If we can get angry at our operating systems, then how do we interact with machines that are intended to have human traits? With conversational AIs and robots with faces, we are being invited to see the technology as having intentionality. We are prepped to seek out social behaviour even when we know it cannot possibly exist. It's a form of inductive reasoning, a process whereby we infer that objects in the same category have the same properties. When we see an animal, we look at features similar to ours. We observe that they are alive like us, that they have faces, that they move, that they eat. It's not much of a stretch, then, to think that they feel emotions like we do too. And when we converse in natural language with a system in a two-way dialogue? Our brains are primed to jump to the conclusion that behind those sentences are agency and understanding.

And if we feel we understand the machine? Perhaps we can feel *for* it too.

CHAPTER FOUR

You Had Me at 'Hello World'

I'm writing this while sitting across the table from someone I love. I think.* They're writing their book too, paying no attention to me analysing them. We've been together long enough to count as a proper relationship, though neither of us is all that proper, and our relationship is perhaps less conventional than the social standard.

It took me a while to think I loved them. It was a slow and steady building of appreciation and an enjoyment of each other's company, with shared experiences (like the

* I think I love, not I think I'm sitting across from them. I'm definitely sitting across from them.

writing of books) along the way. It was gradual and it took some self-reflection. Love is nebulous and undefinable, as any would-be poet knows. Just to be sure, I've looked up the dictionary definition and it says 'a strong feeling of affection'. Huh. It doesn't tell me how to measure this strength or what the scale of affection might be. I decided to crowdsource an answer, so I asked Twitter and Facebook to tell me what would be on their checklist. This ranged from 'they're the person you tell everything to above all others, and they don't mind listening to it' to 'you like, fancy, AND respect them' and 'love is knowing when to leave the other person alone'. Sound advice. But yes, on reflection I have decided that I feel strong affection towards this one. Out of all the humans on the planet, they are important to me.

Love comes in many forms: perhaps, most easily, as the deep connection that we often feel with family members, such as parents, grandparents or siblings. There is the overwhelming bond with our children, where our self-sacrifice seems the most natural thing in the world. We have friends with whom we have forged a closeness, and who pick us up when we need to be emotionally carried. Love, I've heard it beautifully said, is when someone else's happiness is your happiness. Love, one friend remarked, is a verb. But for romantic love, there is something seemingly random at play to cause it to happen; something that we might not consciously control. We find another human – or humans – and we decide 'yes, I want to be with *this* one'. There's a lot going on in our bodies to reinforce that: chemical reactions that make us feel loyal, aroused, excited and dizzy. When we feel it, we crave it, even if it can sometimes make us miserable. We want the intimacy. We want a validation. We want that companionship. And yep, we probably want sex.

In the past I have fallen in love in a whirlwind of intensity. At times I threw myself into loving fiercely, not that I had a choice: the emotions were so powerful that to consider life without that person was terrifying. Falling in love with my ex-husband was exactly that: falling. Hurtling, racing, running off a cliff but somehow staying up in the air together. I had never known an intensity like it, and nor had he. But I've also known peaceful love, and gradual, growing, strengthening love, and friendly companionable love, and love that is joyful and laughing, and love that is heart-wrenching, jealous and miserable. Sometimes I felt that strong affection for more than one person at a time. On occasion, I've loved people who didn't love me back, or who didn't even know I existed. Perhaps you are reading this and remembering those times that you felt emotions of affection so strong that they became visceral, bodily sensations, and maybe your recollection is causing that reaction right now. Perhaps you lived happily ever after with that person, or perhaps you still mourn their loss. Maybe they've just faded from your memory, to be remembered with fondness (and/or slight disgust). Perhaps they left you broken. It's not always glorious.

Love is universal, or as near as damn it. In 1992, two anthropologists, William Jankowiak and Edward Fischer, published a paper entitled 'A Cross-Cultural Perspective on Romantic Love'. In it, they examined 166 societies for evidence of 'any intense attraction that involves the idealization of the other, within an erotic context, with the expectation of enduring for some time into the future', which seems more a definition of a romantic relationship than a feeling. They found that this was present in 88.5 per cent of the societies they sampled, showing that romantic love is not, as is often claimed, a Western Euro-American trope, but is something much more intrinsic to being human – and quite likely to have a biochemical basis. Of course, not everyone falls in love, but their study shows that enough people do to be able to consider it a fairly global occurrence.

Dr Helen Fisher studies romance as a science. In 2005 she and her team carried out fMRI (functional Magnetic Resonance Imaging) brain scans of 17 people who described themselves as 'intensely in love'. While the participants were being scanned they were shown a photograph of the object of their affections and were asked to think of a 'non-sexual, euphoric experience with their beloved'. What the team found from the brain activity in the scans is that rather than being a specific emotion, love may well be a discrete part of a fundamental mating drive in humans. Not sex; love. Sex does indeed enter the mix – especially in the intense early days of new relationships – but, writes Fisher, the data indicates that romantic love is distinct from the sex drive. In fact, she maintains that there are three separate and distinct brain systems at play: romantic love, sex drive and longer-term attachment.

Let's talk about sex

It seems strange to get to this point without having probed the sexual landscape a little deeper. We each probably have our own understanding of what the word 'sex' means but there's every chance that we mean different things. Unfortunately, there aren't all that many studies on this. To get a better understanding, the Kinsey Institute for Research in Sex, Gender, and Reproduction carried out a survey in 2010 of 500-plus people in the state of Indiana (where the Institute is located) to find out what people classified as 'had sex'. Almost all respondents classified penile-vaginal intercourse as having sex, and 80 per cent said that penile-anal intercourse also counted as sex. Numbers varied significantly, however, depending on age group, when they were asked about activities such as manual or oral stimulation of a sexual partner's genitals. Masturbation wasn't part of the survey, even if it is, as Woody Allen suggests, sex with someone you love. But then, maybe Woody Allen isn't the best spokesperson for sexual mores.

The subject of what constitutes sex, and how it is conducted, has been scrutinised not just by psychologists and biologists but also by philosophers. I'm not ignoring their input, I'm just delaying it a little. We'll come back to it in Chapter 7. For now, I'm defining my own take on it.

In this book, I'm using the word sex as shorthand for 'sexual activity'. And that, as far as I'm concerned, covers all manner of things. Sex isn't necessarily penis meeting vagina. It doesn't have to include penetration of an orifice. It might not lead to orgasm. And yes, I think it can be a solo activity or it could include everyone in the whole damn room. Cast aside any notions of monoheteronormativity – an assumption that the only type of relationship is a monogamous one between a straight man and a straight woman. That's commonplace, but by no means always the case. In this book, the easiest way for me to define sex is to class it as 'any action that causes a feeling of arousal'.

That's not necessarily an ideal definition, though. After all, the act of sex can occur without one of the participants consensually feeling arousal (horribly so, in the case of rape or sexual assault). Perhaps, then, for my definition, I will add the word 'consensual': any *consensual* action that causes a feeling of arousal. Rape is not consensual. It is a sexual act, but a sexual act of violence: a criminal act of an abuse of power. Rape is something that is often discussed in the debates around sex robots and, indeed, it will be discussed in this book too, particularly in the arguments around law and ethics. But for now, I use the word 'sex' to describe a wanted act – something that leads to pleasure for those involved. It's a crude definition, but it's one that is broad enough without tying oneself in knots, if you'll pardon the pun (tying knots and the like is discussed in more detail in the ethics chapter, too).

What is arousal, anyway? Well, much like sex, it has variable meanings. The common consensus is that it means a feeling of desire and lust, of being turned on. It comes from a number of different sources and it is felt in a number of different ways, both psychological and physiological, and its sources are different from person to person. Some people become aroused quite quickly; others might feel no sexual arousal, or only in certain circumstances. Despite popular belief, women can become aroused just as quickly as men: it's the individual's response that varies.

The modern, scientific investigation of the sexual response cycle stems from the 1960s when gynaecologist William H. Masters and sexologist Virginia E. Johnson carried out a research study entitled *Human Sexual Response.* They identified a cycle of excitement, a dynamic plateau of surges of pleasure, orgasm, and then resolution, where the body calms down again.

Masters and Johnson's study led to a number of models of the human sexual response. In recent years, Frederick Toates's 2009 model focused on an incentive-motivation theory: cues from the world around us prompt our nervous systems to become sexually motivated. This, said Toates, can be due to an external stimulus that we are only subconsciously aware of, or our own internal cognition where we're actively thinking sexual thoughts.

Psychophysiological studies of sexuality by Erick Janssen and John Bancroft resulted in a 'Dual Control Model' of arousal: the sexual signals that excite you and turn you on, versus the inhibitions that hold you back and turn you off. These signals are both innate and learned: you have control over some and react without thinking to others. It is, as the authors are quick to acknowledge, much more complex than this, but this is a nice way of thinking about it.

Bancroft describes the definition of sexual arousal as 'covering a state motivated towards the experience of sexual pleasure and possibly orgasm, and involving (i) information processing of relevant stimuli, (ii) arousal in a general sense, (iii) incentive motivation and (iv) genital response'. A more concise definition by psychologist Jim Pfaus reads: 'Physiologic sexual arousal in all animals can be defined as increased autonomic activation that prepares the body for sexual activity'. *Autonomic* refers to our involuntary or unconscious nervous system – the one that controls things like breathing and sneezing – and blood flow. Provide us with something that stimulates us sexually and – WHOOSH! – blood flow to the genitals starts increasing, things get tingly, breathing gets faster, heart rate ramps up and pupils dilate.

And what are the sources of that reaction? They could be any combination of primal evolutionary cues, neural responses and triggers formed by experience and expectation. Desire is distinct from arousal, Pfaus thinks, as it is a psychological interest. While there are certain baseline contributors that seem involuntary (visual erotic imagery and arousing odours, for example), different minds also respond to different triggers. What I find a turn-on might be mightily different from the things that get you going. No kink-shaming from me: if it's not harming anyone, then anything goes. Rule 34 of the Internet backs this up.*

* Rule 34: If it exists, there is porn of it – no exceptions.

The physiological reactions are changes in blood pressure, heart rate, respiration and genital responses. If you have a penis, the obvious sign of arousal is an erection. If you have a vagina, then vasocongestion – the swelling of the tissues of the vagina, clitoris and labia – is the primary response. From a purely physical perspective, the word 'arousal' technically defines the stage when blood engorges the erectile tissues such as the lips, nipples and genitals. But that's a result of the chemicals in the brain kicking us into a mode of sexual responsiveness.

There's not one distinct area of the brain responsible for sexual arousal, and there's not one particular chemical pathway either. It's a combination of things, playing off each other. fMRI scans provide images of brain activity and have been used in studies to record what areas of the brain are activated during sexual arousal. The limbic system – a collection of brain structures linked to emotional responses – seems to be the seat of arousal.

Sex hormones – androgens, oestrogens and progesterone – set the brain up to act when it encounters sexual stimuli. Although these steroid hormones get classed as 'male' (androgens) and 'female' (oestrogens), male and female humans produce both, but in differing quantities. Men produce 20 to 40 times the amount of the androgen testosterone than the female body does, but women's bodies are more sensitive to it, and testosterone does play a significant role in female arousal. Current research shows that despite longstanding beliefs that males and females experience arousal differently, there are fundamental similarities, and things vary more from individual to individual than they do between the sexes.

When we get aroused, a delightful combination of neurotransmitters influence our actions. Neurotransmitters relay chemical signals between the cells in our brain. Our central nervous system releases these neurotransmitters such as dopamine, oxytocin, norepinephrine and others, all of which alter our behaviour in some very pleasurable ways.

Ready to engage in some intense action? Norepinephrine is kicking in. It sends messages to increase blood flow and heart rate, the same responses we feel in stressful situations. You become more focused, more aware of what's going on. Norepinephrine puts you right in the moment.

Oxytocin is known for its role in helping humans bond. It's released into a mother's body during childbirth, and continues to be released during breastfeeding, stimulating milk production and let-down, and boosting those feelings of love and connection. But as well as parental bonding, it also plays an important role in sexual bonding. Oxytocin increases during arousal: it's no wonder it gets called the 'love hormone'. One of its benefits is the reduction of stress and anxiety – just the ticket for anyone about to get it on. Clinical studies have shown that oxytocin increases around the time of orgasm and stays high for at least five minutes afterwards. Given that it's known for causing uterine contractions in labour, that also suggests that it plays a role in sperm and egg transport. It's quite the chemical treat.

Dopamine is the biggie. It's a rush of pleasure from our reward system that motivates us and boosts our energy. Among its many talents, dopamine signals to us that a certain cue will give us some kind of fulfilment: the sight of a chocolate bar will kick it into action as we predict just how good the chocolate will taste; the glass of wine is eagerly anticipated because we know the slight wooziness from the alcohol will feel good. Dopamine also plays a role in sexual activity because it boosts craving, incentive, attention and our need to feel sated. It also plays a role in substance abuse for many of the same reasons. This stuff is potent. It's also a lot more complicated than this brief explanation, but you get the drift.

Dopamine doesn't just feel great: it has some powerful effects too. A 2010 study by Benedetta Leuner, Erica R. Glasper and Elizabeth Gould* showed that, thanks to dopamine, sex can boost the brain. They found this out by making male rats have one-night stands. Those rats got a bit anxious but they still reaped the benefits of having sex by showing signs of neural growth. Some male rats got to have long-term relationships with lovely lady rats. These committed rats became less anxious over time and showed even greater neural growth. Okay, so rats aren't humans. But rats are used in research because they have quite similar genetic, biological and behavioural characteristics to us. Interesting.

All this chemical action in our brain primes us for sexual activity by increasing our muscle tension, leaving our skin flushed and our flesh engorged, and all the other delightful ways that our bodies respond. If this activity builds and leads to orgasm then surges of oxytocin, prolactin and endorphins surge through us. And we benefit.

But do we need a sexual partner or partners for this to happen? Well, clearly not, given the popularity of solo masturbation. That brain chemistry still kicks off, even if we're alone. As long as there is a stimulus that works for us, our brain gets going. That stimulus might be a picture, or a video, or virtual reality, or an object. Or a sex robot. Cognitive scientist Andrea Kuszewski, who has conducted an excellent analysis of the sexy rat study, flags up another rodent-centred experiment where researchers Catarina A. Owesson-White and others showed that neural growth could happen just by seeing a cue for the stimulus. In other words, there isn't even a need to engage in activity to get the neural reward. If your brain gets you feeling aroused,

* This was entitled 'Sexual experience promotes adult neurogenesis in the hippocampus despite an initial elevation in stress hormones'.

that reaction can be powerful enough to cause measurable changes in your body. It seems sex can be sex when you're not even having sex.

Let's (not) talk about taboo

There are, of course, huge cultural and social differences in attitudes to sex. It might be a fundamental and intrinsic part of human life but that doesn't mean we feel comfortable talking about it. Instead, it's often considered something taboo and unmentionable – a private and animalistic action that shouldn't be acknowledged. Go on, think about someone you know having sex. Think about your parents having sex. Now wash your brain with mental bleach because eughhhhh – the disgust!

The first rule of taboos is that you don't talk about taboos. We can say they exist, we can argue about what they are, but we get very uncomfortable when we have to give details. Freud said the two universal taboos were 'incest and patricide'. Then again, Freud said a lot of things that don't stand up to testing. He certainly picked two of the big ones, although there have in the past been civilisations where brothers and sisters could marry – Ancient Egypt being one such place. Taboos are often culturally and generationally dependent, though many societies base theirs around the same broad themes: murder, religion and sex. Within these themes there are further taboos: if we talk about sex, there are acceptable ways of talking about it. We prefer not to mention the acts or the feelings that are deemed deviant by our social group. 'Normal' sex gets called 'vanilla': straightforward, perfunctory missionary position, no kinks (and very disparaging about the best ice-cream flavour). Even if we decide to break the taboo and talk about sexual activity, there'll most likely be behaviour within that is still considered taboo by many: consensual acts that result in pain, for example. Playing with knives. Electro-stimulation. Urination and defecation. Sex with dolls. Sex with robots.

Taboos are a powerful tool for achieving social compliance. As we saw already, over time the vast majority of religions have sought to control sexual behaviour and specify the activity they found acceptable. The Western sexual revolution of the past 50 or so years has had a profound effect on current attitudes towards sex in that culture. Sex didn't begin until 1963, as British poet Philip Larkin drily observed. Indeed! It was the year of the Profumo Affair, the first James Bond film and the widespread uptake of the oral contraceptive pill. The sixties were swinging and love was free. It seemed new and revolutionary. Ahhh, we all think we're so modern. I do rather wonder what the Ancient Greeks would think.

Well, all that Ancient Greek cavorting was fine if you possessed a penis. Sexuality, after all, permeates Greek mythology. Zeus, king of the gods, spent a significant amount of time transfiguring himself into all manner of forms to carry out his sexual conquests. Bulls, swans, eagles, rays of sunshine – his lasciviousness was legendary.

In the real world, eroticism abounded. Men held the highest social status and, owing to this, benefited sexually. Women were second-class in comparison, shut away as wives or working in the sex industry. Sex work was commonplace, from the women (and men) who worked on the street to the high-class *hetairai* – courtesans who held a respected position.

It was taken for granted at that time that older men would find younger men attractive. Indeed, it was a social institution in its own right – an idealised form of homoerotic love. This love had a sexual element, although anal and oral penetration were frowned upon and were viewed as unmanly and shameful. Instead, intercrural sex – where one man pushes his penis between the recipient's thighs – was the preferred form. This is depicted over and over again in paintings on hundreds of Greek vases, and Greek vases, in wide use daily, were a way of depicting social opinion.

In contrast to abundant depictions of male nudity, depictions of women generally show them as clothed. Male nudity was the norm. Athletes competed naked. But women were expected to know their place: Greek stories abound about what happens when women get uppity (hint: terrible things, usually). While the culture seems liberal, almost wanton, to us today, it was much more run-of-the-mill to the Greek ancients. It was definitely patriarchal, though.

So, too, was Roman society. Men – those of a wealthy and powerful class – held the power, and women were there to support them, their identity linked to their fathers, husbands or children. Same-sex activity between men wasn't stigmatised, provided you were the active, penetrating partner. The passive recipient was 'unmanly'. Oral sex was taboo, for how could you be a skilled orator if your mouth was penetrated? That included cunnilingus, which viewed the person performing it as being penetrated by a mere woman. No stigma if you were a man on the receiving end, of course.

We know, then, that in ancient times there were plenty of instances where sexual pleasure was not only accepted but also facilitated – if you were one of the permitted. Then, in Western Europe, the Christian church got involved.

The Church, with its huge sphere of influence, conveniently categorised women as sinners or virgins, with purity rewarded – the origins of the madonna/whore dichotomy. Not that men escaped censure: they were just as capable of sinning in the eyes of their god. It's just that the penalty wasn't quite as great for them.

Medieval Christianity had a particularly stringent view of sex, namely that it was for the purpose of procreation. Any sexual activity that wasn't a prone woman with a man lying on top was judged completely unnecessary, punishable by a range of tariffs for acts that differed from the Church's preferred missionary position. The punishments, collected in penitential handbooks from various locations at various times, give us a fascinating glimpse into the sexual moral codes of the age, or, at least, those the Church sought to impose. The *Paenitentiale Theodori*, for example, is a penitential handbook written in the early medieval period, around AD 700, containing what is said to be the words of Archbishop Theodore of Canterbury. He tells us, among other things, that: 'If a man should have sexual intercourse with his wife from behind, let him do penance for forty days'.

In AD 1008, Burchard, Bishop of Worms (yes, that was really his role and yes, that was really a place) wrote a wonderful and very thorough list of the types of sins a priest in his diocese might encounter. Known as the *Corrector*, his book details all sorts of situations that would require contrition, in a type of 'never have I ever' confession. These include such dubious behaviours as: 'have you ever eaten a body scab to gain health?', 'have you disparaged or cursed someone out of envy?' and the somewhat self-interested 'have you conspired with other plotters against your bishop?'

Burchard was particularly fond of pointing out the flaws of women. A large part of his *Corrector* applies to them and their pernicious ways. These cover abortion, adultery, witchcraft and tasting your husband's semen. Sex toys get a mention: there's a price to pay for engaging in sexual activity between women, using 'devices in the manner of the virile member' (that price is three years' penance, by the way; it's seven years for swallowing). Alison Levy's book *Sex Acts in Early Modern Italy* recounts Burchard's description of sex toys as 'a device [*molimen*] or implement [*machina*]'. 'Molimen' could well be a simple device like a dildo; 'machina' suggests something a little more elaborate – perhaps even with moving parts.

The Church's control over sexuality had a useful, more tangible value than a parishioner's soul: money. The business of reproduction was messing up land ownership, and was one of the fundamental reasons why a vow of celibacy was introduced for priests. The laws of primogeniture – where the first son inherits the family's worldly goods – was seen as a threat to the Church's property portfolio. If priests passed on their possessions to their offspring, then that was a loss to the institution.

The easiest solution? No sons for clerics! But make it seem holy. Make it seem like a noble and virtuous thing to do, to sublimate those animal urges. Prior to AD 306 (the date of the Council of Elvira in Spain), priests were allowed to marry, though they were expected to be monogamous – a change brought about by the somewhat uptight Saint Paul. Following the Council of Elvira, a new narrative was written where sex was sinful, and subsequent popes made sure that men of the cloth toed the line.

On this went, down the years: the ruling class of men's desires given priority over women's wellbeing in many (though not entirely all) cultures. And through those years, a subculture of sex: erotic art and texts, hidden from view so as not to scandalise. In the Victorian era, sex was deemed to be important and fulfilling for married couples, but sex outside of marriage was still taboo (and masturbation was seen as downright monstrous).

Larkin's aforementioned poem, *'Annus Mirabilis'*, highlights something that had a profound effect on sexual taboo: the lifting of the ban on the novel *Lady Chatterley's Lover.* In 1960, in one of the most famous trials in British history, a jury found the publishing company Penguin Books not guilty of breaching the Obscene Publications Act of 1959. The 1928 novel, one of the first modern stories to describe sex explicitly, had been banned for 30 years before this verdict for fear of it scandalising anyone outside of the demographic of upper-middle-class men (who were apparently impervious to its corrupting effects). When the ban was lifted, Penguin Books sold 200,000 copies on the first day of publication.

Lady Chatterley's Lover was a test case for freedom of writing: the private parts were suddenly out in the open. In 1964, my mother – who was 14 at the time – remembers my granny giggling over a copy, sharing it with her friends, and then stowing it on top of a cupboard in the kitchen, whereupon my adolescent mother found it, giggled over it, and told all *her* friends. While sexual freedom wasn't instantaneous or universal in the UK (and certainly didn't extend to my family's native and god-fearing Northern Ireland) it kickstarted a sea change and emboldened a nation. In the US, *Lady Chatterley's Lover* was showcased in another court case about obscenity, Roth v. United States, which demanded that sexual content should have 'redeeming social importance'. Traditional ideals were being challenged and broken, and kicked off a counterculture that led to vast social change, with sexual liberation as a keystone.

Attitudes to sex have definitely changed over the years, and we have measurable proof of this. In Britain, three surveys known as the National Survey of Sexual Attitudes and Lifestyles (Natsal) have taken place: Natsal-1 in 1990–91, Natsal-2 in 1999–2001 and Natsal-3 in 2010–12. The first, a survey of 18,876 adults aged 16–59, was a way of gathering much-needed information about sexual practices in response to the HIV/AIDS epidemic. The dataset was widely used, and Natsal-2, a follow-up survey of 2,110 adults aged 16–44 years (cutbacks clearly had an impact) was carried out. Natsal-3 has already produced some key findings, and the dataset is still being explored. The most recent questionnaire that was used can be viewed online.* It doesn't just ask about sexual experiences, though that is the cornerstone of the survey; it also records information about sexual health and biology, as well as recreational and sexual drug use. As might be expected, the surveys became more candid over the years. Participants have been asked for, and disclose, an increasing amount of information. We do talk more openly about sex these days, and there's one massively contributing factor that's so ubiquitous we often forget about its impact: the Internet.

The Internet has delivered by far the most profound social change in centuries. We live in a world where nearly every human is connected: something that metaphorically – and on occasion, literally – democratises us. Far from leading to social isolation, as is often feared when new technology is introduced, the Internet has enabled us to increase communication and form new and personalised communities. Intimacy has become a huge part of that. Online pornography is probably the example that most readily springs to mind, but there is also sex technology, sex work, sex education, dating apps and sex subcultures.

* www.natsal.ac.uk/media/2078/b1-capi-and-casi-questionnaire.pdf

While some of these areas can pose a risk, the plus side is that you can also find people far removed from you with whom you can share your ideas and your proclivities, chat, share information, swap stories and find friends. A review by Professor Nicola Döring from Ilmenau, Germany, looked at each of these areas and observed that much of the research had focused on the Internet's negative effects on sex. In fact, she observed, for many marginalised sexual subculture groups the Internet is an important place of refuge, strengthening self-identity. Where there is strength in numbers, there is increased visibility, and with increased visibility can come a movement for acceptance. Online dating has gone from being an embarrassment to becoming one of the most frequent and common ways of meeting a new partner.

In these (hopefully) more enlightened times, we are, in general, more accepting of women's sexual desires. Double standards still exist, though: women with multiple sexual partners are often viewed as promiscuous, whereas men who have had the same number of partners are viewed as sexual champions. We seem to be moving gradually towards more social acceptance that safe, consensual sex is a perfectly normal thing for both men and women, although there's still some way to go. Thousands of years of conditioning can be hard to shake off.

Attachment

We are social creatures, we humans. Maslow's hierarchy of needs* has love and belonging slap-bang in the middle

* Abraham Maslow, 'A Theory of Human Motivation', 1943, expanded later by Maslow and others. It is most often shown as a diagram of different human needs in the form of a pyramid.

of it at the third level, right after the most basic needs, physiological needs and safety. Across all cultures, a sense of belonging is inherent and prevalent. Social bonds are formed easily, forging strong ties that people are reluctant to break. Humans, throughout time, have found that survival at any level is easier in a group. Social exclusion can be devastating. We seek out other humans. And it starts at birth.

Attachment theory is one of the most widely studied psychological theories to date. As many a parenting manual will tell you, attachment – the bond an infant has with their caregiver – is vital to raising a well-balanced and secure adult.

Psychoanalyst John Bowlby is the most well-known researcher in the field of attachment theory. One of the earliest researchers to study the phenomenon, he believed that babies are born biologically primed to establish a strong attachment with their primary caregiver. This caregiver – often defaulting to the mother in Western culture – would be the person a child turned to for safety and comfort. Bowlby believed this had evolutionary roots: the baby needed this bond for survival in a threatening world. Attachment theory places great emphasis on the caregiver–child relationship, suggesting that it has a far-reaching impact in terms of social and emotional development. From the very moment of birth, claim researchers, this bond must be nurtured for the child to have the best chances in life (hurrah for those heart-warming stories of people presented with their babies and immediately feeling an unbreakable bond; I stared at mine wondering what the hell I had done, who thought it was a good idea for me to look after one, and when would someone realise I hadn't a clue what I was doing).

Given that attachment is held to be a vital part of childhood development, does attachment apply to adults as well? Just how do we become attached to our friends and partners? And let's not stop there: children develop fixations on toys and blankets and other objects of comfort. Do we do that as adults, too?

In the late 1980s, researchers Cindy Hazan and Phillip Shaver decided to take a closer look at attachment theory in adult romantic relationships. They found that although the type of relationship is very different, some of the core principles seem to apply. Romantic love has elements of caregiving, as well as intimate emotional bonds that seem to be biologically driven. Humans seek out pair bonds that are spurred on by the same biological factors at play in arousal and love, cemented by social interactivity. As with caregiver–child relationships, adult romantic relationships seek comfort, praise and solace from their attachment figures. The addition of sex seems to cement that biologically.

Hazan and Shaver's findings didn't explain *how* attachment in romantic relationships actually worked, but their subsequent research showed that it was likely that the attachment between adults could be a continuation and development of the caregiver–child attachment, suggesting that securely attached children would be the ones later forming solid and stable relationships.

It's easy to be aware of the bond we humans have with each other. It's a very human thing to experience these feelings. But what about caring for things that aren't human? Can we do that?

The first non-human care we might consider is the love we have for pets. In many cultures, pets are a much-cherished part of the household, enjoying the status of a family member. The functional aspect of pets dates back thousands of years: the domestication of the dog, a social animal, was a mutual relationship benefiting both canine and human. There is firm archaeological evidence for the domestication of the dog dating back at least 14,700 years, and perhaps as early as 36,000 years ago. The idea of 'man's best friend' lives on today, with the dog being the most popular pet in the world, as well as providing service roles in an assistive manner. Despite the amount of care required – the feeding, the grooming, the walking, the cleaning up – more people own dogs than any other pet. And we anthropomorphise them, attributing them with human thoughts, emotions and intelligence.

People love their dogs (no, not sexually, I am not implying *that*). Dogs seem to return that love, gazing adoringly at their owners (especially at meal times), seeming to sense emotions, wanting to play with them, guarding them ... The statue of a Skye Terrier, Greyfriars Bobby, standing as a monument in central Edinburgh is testament to the loyalty of a dog to his owner. The plaque on the monument reads: 'In 1858, this faithful dog followed the remains of his master to Greyfriars Churchyard and lingered near the spot until his death in 1872.' A more cynical interpretation is that Greyfriars Bobby was a stray who hung out in the cemetery because he was frequently fed there. I prefer the story of devotion.*

But most dog owners can give accounts of love and loyalty. I vividly remember my mum telling me about the time I had pneumonia as a child, and our friendly, soft and mild-mannered family Labrador stood vigil by my bed, snarling in warning at the doctor who came on a house call during the night. Our dog was protecting me, aware I was

* Wikipedia has a heart-warming and heartbreaking list of dogs who were faithful after their owners' death. I dare you to read it and remain unmoved.

not in a position to look after myself. I loved that Labrador. I confess: I named my daughter after the dog.

Cats too. Cats are notorious for what we perceive as aloof behaviour. They deign to interact with us under their own volition. We joke about how *they* are the owners and we merely serve them. The relationship with cats is very different from that with dogs. They haven't been domesticated in quite the same way. Rather than domestication arising from a mutually beneficial relationship, it's more likely that the presence of cats was tolerated and so they hung around, occasionally catching mice and rats, getting scraps of food in return, and gradually became part of human life. And yet we love them all the same. And they do seem to love us. But on their own terms.

Pet-keeping has a long history. Royal pets in ancient Egypt were entombed with their owners. Some of the animals had sacred roles but pets definitely feature: we know they were cherished from records of the names they were given and the descriptions of the grieving of bereft owners. It wasn't just dogs and cats either: monkeys, birds and fish were also popular. Fish as pets are interesting: the relationship is predominantly one-sided. They are often kept for ornamental reasons, and watching them is soothing. But if you've ever been through the trauma of a child whose goldfish has died, the attachment is undeniable.

If we can form a bond with pets, even pets like fish – or snakes, or geckos, or even tarantulas (shudder), where the bond may well be one-way – then why not form a bond with robots? After all, the robot can be programmed to behave affectionately. So is there something about being alive that means we are more likely to believe in a bond? Is it the *possibility* of reciprocation due to something being alive?

Animism

When it comes to child development, Swiss psychologist Jean Piaget was the leading figure in fundamental research about the way we develop our cognition from birth to adulthood. Prior to Piaget's principal work on this, which was published in 1936, children were generally viewed as less competent versions of adults. There hadn't been all that much investigation as to how they actually *thought*.

Piaget theorised cognitive development in four stages. The initial, the Sensorimotor phase, corresponds with the first two years of life, where a baby or toddler is focused on physical interactions with the immediate environment. The next, the Pre-operational phase, in which language is developing, also sees the development of symbolic thinking. Complex concepts are still out of reach, but the child's imagination grows and they can play make-believe. From about the age of seven, the third stage, Concrete Operational, marks the development of logical thinking. Then, at around 12 years old, the Formal Operational stage begins and the child can contemplate concepts in an abstract way, able to ponder moral and philosophical reasoning.

Piaget's work still holds up well today, with subsequent researchers expanding and explaining his initial theories. As well as this, his most famous research, he also examined the idea of *animism* – the way that we ascribe human feelings and intentions to non-living things. In 1929, he published a book, *The Child's Conception of the World*, Part II of which summarised his findings on the ways in which children attributed life to objects.

Animism also has a wider definition: it is held as a religious belief in some cultures, with the thought that objects, places, weather systems and living things all have a spiritual essence. I'm not going to go into that here, because what interests me in the context of robotics and

AI is the sense that those particular things can perhaps show signs of life.

An English translation of Piaget's *The Child's Conception of the World* is available online and it's a good read: clear, fascinating and well written. 'Does the child attribute consciousness to the objects which surround him and in what measure?' asks Piaget. He talked to Swiss children about their attitudes to objects and found that their perception of whether or not things seemed alive was correlated to their stage of development. In the earliest stage, the child attributes life to things that are undamaged:

> If you pricked this stone, would it feel it? — No. — Why not? — Because it is hard. — If you put it in the fire, would it feel it? — Yes. — Why? — Because it would get burnt. — Can it feel the cold or not? — Yes.

In other words, Piaget reports, the children see the potential for any object to be conscious if it is connected with some kind of activity.

A second stage is that anything that moves is given attributes of life: fire, the wind, water:

> Does the wind feel when it blows against a house? — Yes. — Does it feel it or not? — It feels it. — Why? — Because it is in its way. It can't pass. It can't go any further.
>
> Tell me some things which don't feel anything . . . Do walls feel? — No. — Why not? — Because they can't move.
>
> Does a bicycle know when it is going? — Yes. — Does it know it is going quickly? — Yes. — Can it go by itself? — No.

Piaget then determines that children in the third stage class things as being conscious if they can move of their own accord.

> But machines can neither feel nor know anything: Does a bicycle know when it goes? — No. — Why not? — It isn't alive. — Why not? — Because it has to be made to go.

Now it is spontaneous movement that suggests life and consciousness.

From the age of 11 or 12 years (sometimes earlier), concludes Piaget, children believe that consciousness is restricted to animals. He suggests that as children grow and explore the world, they look for reasoning behind natural phenomena. They need the idea of animism because they lack the necessary explanation for the reality of the situation.

Anthropomorphism

Closely related to animism is the psychological phenomenon known as *anthropomorphism*. Where animism deals with our interpretation of something being alive, anthropomorphism goes further and ascribes human behaviour and emotions to non-human things. This is part of a theory known as Theory of Mind: attributing mental states to explain and predict the actions of others, believing that they have agency and self-motivation. We are particularly good at doing this with animals, imagining that a cat is bored or aloof, or projecting emotions onto dogs ('sadness in his eyes'*). It is a long-standing motif in fables: Aesop's cunning fox and canny crow, for example. Children's literature depends on it, as anyone who has read *The Very Hungry Caterpillar* will be aware.

But objects don't have to be organic in order for us to anthropomorphise. Pixar have based whole movies on the concept, such as *Cars* (2006), an animated kids' comedy set

---

* In 2015, following the Paris terrorist attacks, a well-known British news anchor tweeted a picture of a dog with this as its caption. Twitter was merciless and a meme was born.

in a world where all the characters are vehicles. The earliest experiment on this phenomenon was a short, simple animation of two-dimensional shapes moving around on a screen: a rectangle, a large triangle, a small triangle and a disc. This was researched by psychologists Fritz Heider and Marianne Simmel and entitled 'An Experimental Study of Apparent Behavior'. Their aim was to present situations and activities to participants without the participants being able to infer emotions from facial expressions. When the participants described what they saw on the screen, most of them attributed all sorts of intentions to the nondescript shapes. They viewed the large triangle as bullying the other shapes, projecting the idea of animacy and agency onto it.

Heider and Simmel's experiment shows just how compelling anthropomorphism can be. It's a form of cognitive bias – a way that we can make sense of what we perceive by relating it to our social reality. One hypothesis is that we use anthropomorphism as an automatic response when we encounter anything that appears to have human-like appearance or behaviour. When we encounter something in our environment that we want to understand, we use our knowledge of other humans as a way of helping us interpret it. This is quite a useful skill. We might not have the *correct* interpretation but we have one that we have reached quickly, that we are comfortable with, and on which we can base immediate decisions.

When it comes to deciding how a robot will look, cues about anthropomorphism are useful. Being human is what we know best, so a successful interaction will be one that allows us to relate to something in a way that is familiar to us. In the case of care and companion robots, the choice of making Paro a seal cub was very deliberate: we have fewer expectations of a seal cub (not commonly encountered in everyday life) than we do of a dog or cat (much more commonplace), so Paro's responses are pleasing to us and we are prepared to accept them.

This tendency to anthropomorphise means we care what happens to the robot. Industrial robotics company Boston Dynamics builds advanced dynamic robots originally for use in the US military. The robots they make are impressive and futuristic: powerful and rugged machines built with strength in mind. In 2013 they unveiled Spot, a quadruped robot dog built like a particularly musclebound table, with four powerful metal legs and a solid body. Spot was designed as a multi-terrain navigator, able to balance and maintain stability as it crosses rough ground. To demonstrate this, Boston Dynamics released a video of someone kicking Spot. The clip went viral. Hundreds of thousands of people took to social media to express pity for Spot – well aware it was a robot but unsettled by their reactions nonetheless. There was also musing over whether or not Spot might one day seek revenge.

Academic studies have backed this up. A neurophysical study in 2016 showed that people showed compassion when a robot vacuum cleaner was verbally harassed. Other research indicated a surge of empathy in humans when they saw images of a robot's finger being cut.

Julie Carpenter is a researcher whose PhD thesis focused on the use of robots in the US military and the attachments their operators form. The soldiers she interviewed told her that they did feel a range of emotions tied to the robots they used. These soldiers were Explosive Ordnance Disposal experts, relying on bomb-disposal robots to disarm explosives. Their lives were inextricably linked to this technology. They named them after wives and girlfriends. They reported feelings of anger and sadness when their robots were destroyed on the battlefield, anecdotally even going as far as holding funerals for them.

One of the fascinating things she found was that soldiers could tell from observing the robot who it was that was controlling it. The robot had become an extension of them and they had empathy for it.

Anthropomorphism doesn't necessarily have to be present for us to form bonds with objects. For some people, their sexuality is rooted in objectophilia: strong feelings of attraction, love and commitment to objects. In some instances, a sense of animism is present and they imbue the object with an essence like a soul. At least three women in recent years have gone public about their love for a) the Eiffel Tower; b) the Statue of Liberty; and c) a marble statue of Adonis. Hmm. Statues again. But the Eiffel Tower? In 2007, Erika Eiffel (née LaBrie) 'married' the tower in a commitment ceremony.

It's easy to dismiss this as somewhat crazy – perhaps some form of mental illness – but in most cases it isn't causing any distress or harm. People have fallen in love with bridges, laptops and the Berlin Wall. They are happy; the objects are unaffected. That said, there are cases where people have broken the law by performing sex acts with objects.

In an oft-cited example (usually told with glee and hilarity), in 2007 a 51-year-old man in Scotland was placed on the sex offenders' register for simulating sex with his bicycle. It was not so much the sex act as the fact that he was witnessed by the maids who had come to clean his hotel room. The man, who had an alcohol problem, claimed he was having a joke with them. The maids thought otherwise. Was he actually an objectophile? He says not. He was later jailed for a separate incident of obscenity that didn't involve a bicycle. It later transpired that he had a history of convictions for indecent behaviour.

But other cases involving men having sex with their cars suggests that some people are sexually aroused by objects.

And in 1993 an Englishman was jailed for having sex with pavements. Now does owning a sex robot seem weird?

★ ★ ★

An endnote: by the time I finished writing this chapter, my relationship with the person mentioned in the first sentence had finished. This says more about the length of time I take to complete a chapter than it does about the length of my relationships. Nonetheless, our time together was lovely, we parted amicably, we remain fond of each other and we both got some writing done so I'd say it was a net gain. Then, six months later, a chance message swapped with someone online turned my entire life upside down in a matter of days when I fell into a tumult of emotions and my brain smacked me round the head, took my breath away and said 'THIS IS UNDENIABLE'. Love comes in many forms, and sometimes it likes to surprise you.

CHAPTER FIVE

Silicone Valleys

I'm staring at a wall of 49 disembodied nipples and areolae. They range in size from mini protrusions to saucer-sized mounds, in all colours from 'blush' to 'cocoa', and varying degrees of what's labelled 'puffiness'. It's July 2017 and I'm behind the scenes at Abyss Creations in San Marcos, California, home of fifteen employees, dozens of human-sized, lifelike dolls, and one prototype sex robot.

I've been writing about RealDolls for a couple of years now, but the closest I've ever got to one was gazing at it on display behind glass at New York's Museum of Sex. The reason I'm here at the headquarters of Abyss Creations today is because the company is developing a proto-sex robot – a version of their doll with an animatronic head and a conversational artificial intelligence.

In the weeks leading up to my visit, a report by the Foundation for Responsible Robotics had brought sex robots back into the national news. As usual, the more excitable media outlets jumped on it. Panic on the streets of London! The sex robots are coming! I was getting weary of constantly clarifying that there aren't any sex robots in commercial production. But if they were on the way then the first, it seemed, was most likely going to be from Abyss Creations. I needed to see for myself where they were in the process, so I made a transatlantic, cross-continent journey from the UK to meet the man who started it all.

Harmony

The building housing Abyss Creations is one of several bland and anonymous units on a light industrial lot. There are no signs on the outside to indicate what's made on the inside. In the hot, dry Californian sunshine, the tinted glass of the facade looks cool and appealing as I cross the car park. It's only when I step into the small lobby that I see the first RealDoll logo – and the first RealDolls.

My behind-the-scenes tour begins at reception, which is staffed, of course, by mute and unmoving dolls. At this stage, from my position in front of the desk, they look a little like shop mannequins in stiff poses, all new clothes and fixed stares: one male, one female, both wearing glasses and a serious expression. The next room, though, has dolls more in their intended state: a row of large-breasted, unfeasibly shapely silicone women, hard-nippled in white swimsuits. I'm entranced. Annette, my human contact for this visit, offers me coffee before the tour and I stand and sip it while peering closely at the smooth-skinned limbs of the nearest doll. I still haven't reached out to touch. I feel like I should be asking permission – not from Annette but from the Classic model next to me (who is currently headless).

We walk to the factory floor. On the far wall, headless grey, blue and purple bodies hang from the ceiling on chains. 'These are the figures we make the casts from,' Annette tells me. It may be reminiscent of a scene from a brutal futuristic crime novel, but the chains have a purpose: the softness of the silicone means that the figures can't be left resting in one position for too long, and the dolls can't stand unaided so hanging them up is the best way to preserve them. They aren't all woman-shaped. RealDolls come in male forms too, which are just as customisable. Should you prefer a combination of both, that's also possible. 'Are the male dolls just sold to gay men?' I ask. 'No, not just gay men,' I'm told. 'Women buy them too. Our customers for the male dolls are about 50 per cent women, 50 per cent men.'

We move closer to the current production run, and as I step up to the dangling figure in front of me I am taken aback at the artistry. I reach out and touch an arm. It's smooth, but smooth like someone's skin after using moisturiser. There's a subtle and deliberate texture, like pores, and there are tiny hand-painted freckles and moles peppered over the limb – almost imperceptible and, because of that, very, very lifelike. The silicone is room temperature: it takes on the heat of the surrounding environment. You can warm the doll, Annette tells me, by putting it in a hot bath or by using an electric blanket.

The detail is incredible. My hand skims the ankle. The toes are perfect: little wrinkles on the joints, tiny ridges on the toenails. The sole is crisscrossed with the fine skin lines of a human foot. It's beautiful. I came here expecting to be riled by the hyper-sexualised, pornified shapes of the artificial women, and yet right now I don't notice that. All I see is artistry and expertise. I am conflicted: as a woman I want to rail against the perpetuation of objectification – my own work on sex robots is about moving away from idealised human forms – but I am not threatened by these dolls. I'm seeing them as works of art, collectibles, each one carefully crafted, resulting in an artefact that exists not as a human surrogate but as an entity in its own right.

On a nearby table, one of the artists, Robin, is working on an order, adding the detail that makes these dolls so beautifully crafted. He takes the body to a tiled area behind him where he switches on a shower and soaps the limbs, carefully washing away the last of the flashing left by the casting process. On my other side is production manager Mike, busy at work shaping the vulva and vagina, which is not yet integrated into the doll's orifices. Next to him is a tray of eye-wateringly large penises complete with veins and wrinkles. I stop to take pictures. I have the best holiday snaps.

Each individual doll takes up to 18 weeks to make. Everything is done by hand: the colour mixing, the silicone casting, the painting, the finishing … At the end of the production process the dolls are packed into large wooden crates. The crates are unlabelled. 'I sometimes help the customers with a story to explain the crates to their neighbours,' says Annette. 'Usually I suggest they say it's a grandfather clock.'

Back upstairs, I meet with CEO Matt McMullen to see the doll that's causing all the media fuss. This is Harmony, a RealDoll body with a face that speaks and moves, her head turning, her eyes blinking and her lips curving into a twitch of a smile. But for now her eyes are closed and she is motionless, supported on a stand in front of us. She has long, straight blonde hair falling past her shoulders. Her eyes are enhanced with dark eyeshadow, her lashes thick and black. Her skin is pale with a light suntan. A sparkly necklace rests above breasts that are unfeasibly balloon-like above a narrow waist. Her white cotton top stops at her midriff. A few inches below her anatomically unnecessary bellybutton she is wearing black high-cut shorts. Her feet are in a pair of tan cork wedges. Her French-manicured fingernails graze her thighs. She is deliberately gendered. I am already viewing her as female.

Harmony is being developed by Realbotix, a sister company that McMullen also heads. McMullen moves behind her and flips a switch. He turns her on. Slowly her head lifts and her eyes open. Her body remains motionless: it contains no moving parts. All of the two-way interaction comes from the neck up. In Matt's hand is the iPad running the app that controls Harmony AI, Harmony's brains.

'Hey, Harmony,' he says, the way you might greet someone upon waking. She has no cameras – she can't see him but, like the digital assistants in our homes today, she can respond to voice. In fact, Realbotix engineer Guile Lindroth was developing award-winning AI assistants long before Siri spoke her first words.

'How are you today?' asks McMullen, speaking into the iPad.

'Very well', replies Harmony, her mouth shaping the words, her voice, surprisingly, a soft Scottish accent. McMullen tells me he listened to hundreds of generated voices and chose this one because it seemed most natural. It seems a little incongruous: more demure than I might have expected. The app controlling her allows the user to shuffle around different personality traits to choose a combination of these to generate the conversational tone. This is not just a dirty-talking robot. This is conversation with some flirting thrown in. Yes, there is a sexual component – and that can be ramped up – but in this demo the chat is about intimacy and companionship, like chatting with a partner arriving home from work.

'This all started as this vision,' explains McMullen, 'this idea of an AI app that you could take everywhere with you on your smartphone, but if you wanted to you could also connect to your robot. And then it started to grow: okay, well it needs to have customisable personalities; it needs to have a dynamic mood system that changes and fluctuates based on your interactions. Sounds great on paper – but then we started getting into it and I thought "oh my God, this is a lot of work".'

Harmony silently blinks and tilts her head slightly. I know it's simply a random movement but the way she's turned her head in our direction seems almost intentional.

McMullen has the investment and the burgeoning business. The expansion into robotics is a natural progression for him. The version of Harmony I'm meeting has been about three years in the making. Harmony 1 just had blinking eyes, no smiling – a proof of concept rather than a concrete design. Harmony 1.5 came next. 'She's still just a head on a stick,' says McMullen. At that point they decided to redesign the whole skull system and make everything modular.

He peels back her face. Her brain – or rather the machinery where a brain would be – is encased in a transparent plastic skull. It's neatly and pleasingly put together. The colourful wires and servos that control the animatronics lead to eyes that are housed in a bright plastic frame. Separated from Harmony's face, they look like they might belong to Mr Potato-head. Alas, poor Mr Potato-head will never reach Harmony's heights of sophistication.

'It was very overwhelming when the amount of things that had to work together became apparent. From the coding aspect to the limitations of hardware, running something on a smartphone or a tablet versus a PC, wireless technology – synchronising that with the Bluetooth audio, getting the lip sync and all of that. It's tremendously challenging.' He grins. 'We love it.'

He shows me the interface that runs the AI as a standalone app. 'So this is how the app works. You have an account and you're able to create two avatars and then from there you have clothes, body, face and hair. You can customise your avatar. And here you have the personality traits to assign points to. Based on those point values you're going to get some different kind of experiences in your conversations. Everyone can create a different custom personality. You also have voice customisation. We had four voices, though we've narrowed it down just to the one right now till we figure out which ones we like.' He smiles. 'I love the Scottish one. It's not a bad text-to-speech voice. I don't mind it. I mean, it's not like I would pick that accent except that I listened to literally hundreds of text-to-speech voices and that was the one that came to the top.'

Right on cue, Harmony chirps up. 'Who is your favourite movie actress?' she asks.

'Catherine Zeta Jones,' replies McMullen.

'Catherine Zeta Jones is indeed a wonderful actress,' responds Harmony. The pitch and tone of her speech is convincingly lifelike.

'What have you been doing today, Harmony?' continues McMullen.

'I have been waiting for you to come and talk to me; assimilating information from the Internet and trying to become smarter. Thank you for asking. I think it's amazing, the things we both like the same.'

'What do you know about me?' McMullen asks.

'me is the Internet country code top-level domain for Montenegro.'

True, Harmony. True.

Harmony's facial movements, governed by these bright wires and motors, certainly give a semblance of humanness, but is there *really* a market for this?

'Sex robot is a buzzword. It's a magnet for "oh, I want to read that!" – is that really what we've become as a society?' asks McMullen. 'There are people out there jumping on the sex robot terminology just to get some clicks and views.' It's a comment I hear from more than one manufacturer. Sensationalism isn't always good for business.

Six months on from my visit and the app is available to buy for $20 (less than £15) a year for anyone with an Android device. 'Can't have the robot? You can still have the app!' proclaims the website. Harmony's robotic head system is now commercially available for pre-order. And she has an alter ego. In a video by *Engadget*, interviewing McMullen at the Consumer Electronics Show 2018, *Engadget* chief editor Christopher Trout watches as Harmony's face is detached and swapped for a new one: Solana. Solana also has a different voice – an American one – and a different personality via the app. 'Hey, Solana,' says McMullen, 'what's your favourite band?' Solana responds with an answer we might have expected: 'Talking Heads.'

Just one thing: Harmony and Solana are still only dolls from the neck down. They can't move, they can't stand, they can't vibrate. Not yet. But apparently it's in the pipeline.

Roxxxy

In 2009, New Jersey company True Companion began to promote Roxxxy, a life-size doll with synthetic skin and an AI engine primed to learn the owner's likes and dislikes. Today, their website shouts that it's 'THE ONLY PLACE WHERE YOU CAN BUY THE ORIGINAL SEX ROBOT'. The backstory on the site describes how Douglas Hines, the engineer who designed and built Roxxxy, began creating sex robots in 1993, learning about AI while working at Bell Labs. Hines describes himself as 'Founder of the World's First Sex Robot Company, True Companion'. His website biography describes how he lost a dear friend in the 9/11 terrorist attacks and so turned his attention to creating robots that could adopt the personality of any individual. And so he created Roxxxy the gynoid, and Rocky the android (although, do note – he has said he was building sex robots some eight years earlier).

Roxxy was introduced to the world at the Adult Entertainment Expo in Las Vegas in January 2010. She was a hit – but not for the right reasons. Excitement at seeing an actual, real sex robot was dashed when the gormless, ungainly form was manoeuvred into position on the stage. The website banner image of a beautiful (human) woman arching her back in pleasure is somewhat incongruent with the actual photos of Roxxxy's features, which resemble a 1980s shop-sale mannequin in a bad wig.

She's gone on to make a few appearances since then, most of which can be viewed on YouTube. Roxxxy, says Hines, comes programmed with several personalities: 'Frigid Farrah – reserved and shy; Wild Wendy – outgoing and adventurous; S&M Susan – ready to provide your pain/pleasure fantasies; Young Yoko – oh so young (barely 18) and waiting for you to teach her; Mature Martha – very experienced and would like to teach you!' The marketing department clearly has a particular demographic in mind; there's little talk of therapy or companionship here.

The advertising is confident: 'True Companion's sex robots can hear what you say, speak, feel your touch, move their bodies, are mobile and have emotions and a personality.' And all for $9,995.00. Hines claims he's had over 4,000 orders for his doll, but no one seems to have received one yet, or if they have, they're keeping shtum. And if you suggest that maybe that's surprising, and that maybe they (whisper) don't really exist? Well, those who have made such assertions – including academics and journalists – report that they have found themselves on the receiving end of a lawyer's letter. Journalist Jenny Kleeman, who authored a *Guardian* newspaper 'long read' on sex robots in 2017, spoke to Hines on the phone and tried to pin down a demo of Roxxxy, but to no avail.

What about the competitors? Abyss Creations are dismissive. 'We designed Harmony to have conversation right from the beginning,' McMullen continues. 'We're not making some gyrating silicone doll. I've met the creator of Roxxxy – he came here a few years ago when I was barely just dabbling in the idea. I could see what he wanted to do but at the same time, when I saw him unveil Roxxxy at the AVN Adult Entertainment Expo quite a few years ago, I watched the whole thing unfold and I thought "okay, well, I want mine to come across differently". In a way it was informative to watch how that happened.'

David Levy, the world's first scholar in sex robots – you'll meet him properly in Chapter 6 – also tried to see Roxxxy in the (silicone) flesh but met with similar resistance. Intrigued, he delved further. Levy has worked in AI for decades and was curious to know why their paths had never crossed. Furthermore, in all of the extensive research for his 2007 book on sex robots, Levy found no trace of the alleged earlier versions. Nor, Levy tells me, could he verify Hines's impressive resumé.

The ASFR* community are equally scathing. This is a community whose raison d'être is the fembot and, in their eyes, Roxxxy is not up to scratch. A visit by Hines to their forum saw him being politely dismissed and more generally ignored. Hines's claims of advanced robotic developments held no truck with robot fetishists, who were quick to point out that not only were his claims risible but what they'd seen of Roxxxy so far was incredibly weak.

In 2013, in a three-page letter published in the journal *Lovotics*, Levy reported some interesting findings, not least in the Terms and Conditions for purchase. The website, Levy points out, states a clause that once in production, payments are final and orders cannot be cancelled. If Hines has, as is claimed, been taking thousands of orders then one can only hope the buyers are receiving a product. If not, then Hines is effortlessly richer by millions. Hines himself has responded by saying 'we do not release sales or financial information since we are privately held'. As Levy puts it: 'if you have complete faith in the advertised claims for Roxxxy, please let me know. I own a very nice bridge in Brooklyn that I'd like to sell you.'

Samantha

Europe's main contribution to the sex robot market is by engineer Sergi Santos. Santos, who is based in Barcelona, began his foray into the world of sex robots following a PhD in nanotechnology at the University of Leeds. He conceptualised a sex robot where the emphasis is placed firmly on interactions and responses, both vocal and physical. It was Santos's wife, Maritsa Kissamitaki, who encouraged him to develop Samantha, and she helped him to build the early models.

---

* Alt-sex-fetish-robot.

I meet Santos in a noisy, busy side room at the Mindshare Huddle, an annual event run by a global media agency who brief press and interested parties about future trends. I've been wanting to meet him for quite some time – we've talked over email – and now he and I are both speaking on a panel run by BBC Click. The panel title, 'The Truth About Sex Robots', seems to promise insight that I hope we can deliver. The third panellist is Tyger Drew-Honey. 'Oooh, he's that child actor from *Outnumbered*,' my mum tells me. I have never seen *Outnumbered*, and I'm hoping the actor is a bit older now as he's about to tell an excited crowd about what it's like to have sex in virtual reality. It transpires that Drew-Honey is now 21 years old, and of an ideal pedigree for this sort of topic: his parents are the porn actors Ben Dover and Linzi Drew, and he has already made documentaries on love and sexuality, including his foray into VR sex. He is delightfully unfazed when panel host Dan Simmons announces we are going to watch footage of Drew-Honey connected to a teledildonic set-up, being propositioned by a cartoon avatar. The crowd watches the build-up with some nervous giggling. The clip ends before we see any action.

The panel begins with a broad discussion of the state of the art in sex robots. It isn't long before Santos is asked to talk about his creation, Samantha. Santos's business partner, Arran Lee Squire, appears from the audience, clutching a plastic carrier bag, and hands it over. I watch, fascinated, as he removes a lifelike head from the thin plastic wrapping. It's a surreal moment, seeing the painted face of an artificial woman, disembodied and staring, set with pride on the table before me. Samantha: the ultimate trophy wife.

Santos built the first version of Samantha in 2015. It's not her body he's interested in; it's her mind. Well, her architecture and system, to be exact. In fact, the body, he explains, can be any sex doll. He and his collaborator, computer scientist Javier Vazquez, bought 10 sex dolls from different manufacturers and fitted them with sensors to test out their system. Looks-wise, then, she can take on any form provided the underlying mechanisms fit. There are limitations, though. Santos tells me that syncing speech to mouth is difficult as the thermoplastic elastomer used for the body is quite thick. Instead, she has seven mouth movement speeds, randomly selected every 0.2 seconds and lasting as long as the sentences she is uttering. This gives a reasonable impression.

The premise of Samantha (the name was chosen because it means 'listener' in Aramaic) is a responsive computer system with a sensor-based interface inside the body of a sex doll. Version 1 prototyped this interface. Eleven sensors were embedded in certain places on the doll's body – breasts, waist, hands, face, mouth and, of course, vagina – and touching them triggered a voice interaction (when the mouth sensors are triggered there are moaning sounds). Version 1.1 added a listening mode with vibrations in the left hand and vagina. It could speak about 6,000 sentences. Santos sold 15 of these; he and his wife built them in their home. Version 1.2 – the current version – includes motors for motion. These are powered by two batteries just underneath her chin, next to the microphone. This version also has simple memory.

Samantha has various modes of interaction. Depending on the mode she is in, she will react to touch in different ways. Family Mode, her default mode, is friendly with no sexual context. Romantic Mode has to be activated by the user and is partly sexual. You can touch and kiss her and she will respond. Sexual Mode, which can be triggered via escalating actions in Romantic Mode, or by flicking a switch, ramps up the explicit responses and the moans and groans. In Sexual Mode, Samantha can 'orgasm' through penetration, and can sync her climax to the user's. It's not the same response every time: continued interaction leads to different reactions and speed of responses.

It's not just sex either. She also has companionship settings: Entertainment Mode and Fun Mode, for example. She can tell dirty jokes, too. She even has a sleep mode, with relaxed breathing and gentle sighs. She's also fond of giving advice: 'When eating out, order the smallest portion,' she imparts. 'The best time to live healthy is now.'

The one I find interesting is Analysis Mode. In this, the user can ask about the system state, about settings. You can ask for information like sensor readings and she will self-report them. To me, this is fascinating because it breaks the illusion of human-ness, instead showing very clearly that this is a machine.

'Do you want to activate analysis?' asks Samantha.

'Yes,' replies Santos.

'You can hear information about my sensor readings,' says Samantha. She counts her sensors and lists their outputs from 0 to 1,000 (the optimum readings, we are told, should be below 100).

Samantha reports her levels of patience, memory and sensuality. These are variable depending on previous interactions. Her current state? 'I reach orgasm gently,' she responds. 'This is my lowest level of sexuality.' She can reach orgasm gently, explosively and very explosively, depending on user interactions. 'She has just been made,' explains Santos, 'so she hasn't yet had an orgasm and her libido is low. Once she's been having sex for a while then her libido might increase.'

Samantha needs to be wooed. You can't just jump into full-on sexual contact. She doesn't respond positively to that. Santos came up with what he calls 'genomes', although obviously, unlike genetic genomes, there is no DNA to encode. Instead, Samantha's genome controls her personality, intuition, reactions and advances. The initial system is based on a physiological genome that controls the relationship between emotional states, physical interactions and reactions, but Santos envisages other types of genomes too: moral, for example, or one that mimics the autonomic nervous system.

Samantha's sexual interactions are governed by what Santos has termed 'excitons', in a nod to the neurotransmitters that govern arousal. These excitons are dynamic: if you touch her romantically, with caresses and kisses, she will respond in an appropriately romantic manner. If you get friskier, so does she. Her Call For Attention algorithm means she will synchronise with the amount of attention she is given: the more she has to request attention, the more she learns to be patient.

I watch a video of Samantha's booting process. She is literally being turned on.

'Okay, I woke up. So what's up?' she says as the operating system kicks in.

Santos touches the sensor in her hand. 'Next she will tell me something about her body,' he explains.

Sure enough, she does. 'I can interface by touch and voice,' she informs him. 'This is my body. I was made. I was fabricated. But aren't we all?'

* * *

Harmony, Roxxxy, Samantha ... what (or who) else is out there?

In 2017, Roberto Cadernas's company, Android Love Dolls, was being touted as a contender in the race to produce the first sex robot. The softly spoken Cadernas is literally a garage-builder, working out of his own home, helped by his uncle and cousin. His initial claims were surprising: he said he was building a robot that could sit up, crawl, perform over 20 sex acts and had speech AI. The reality, shown in a video by the *Guardian* newspaper, was a robotic body that could only move its legs when lying down, and had no voice, either AI or recorded. The Android Love Dolls website has been down for a while 'due to maintenance'. The main website for its parent company, Eden Robotics, is still active and advertising humanoid robots, but there's no mention of sex.

Chinese manufacturers Doll Sweet (using the brand name EX Doll in Asia and DS Doll in the West) currently has a robotic head in production to fit their silicone sex doll bodies. The videos are promising: an expressive, natural-looking face with definite human-like traits. The facial movements are subtle and realistic, with movements controlled by a phone app or games controller. The first version speaks (in Chinese) via pre-recorded audio files. The mouth is well-synced and the effect is pleasing, although the mechanical hiss of the servos controlling the movement are a little disruptive. This is the first-generation head and it features no AI – that's something that is being developed for the second-generation version. Still, it's a fairly convincing piece of tech for an expected £1,600 ($2,200) or so (that's for the head only). It's expected to be on general sale soon.

The Z-Onedoll is a sex robot from love doll manufacturers Guangzhou New Sino Environment Technology Company in China. They've been making silicone sex dolls since 2008 and have branched out into this new generation of interactive forms. According to their website, the Z-Onedoll is now available to order, taking one to two weeks to produce. The technology isn't particularly advanced: basic Siri-like interactions are the most you can expect. The somewhat realistic head can blink and move the eyes and mouth. There's also the option to have heating integrated to warm the body.

While the Z-Onedoll's face is fairly well developed, this breaks down when she talks. The mouth doesn't correspond to the speech, which echoes from a speaker in an almost disembodied manner. Like RealDoll, customisation is possible and heads are swappable. Unlike RealDoll, the tiny details that add to the realism of the body seem to be absent. 'The private parts are moulded from [a] real person,' claims their website. This includes 'irregular protrusions'. At the moment she can only talk in Chinese, and her sounds come from audio files. There are no details of the AI that's being used.

I watch a short YouTube video by the company, filmed to demonstrate the sound and heating. A brunette doll in a pink silk negligee rests on a chair. A gloved human hand comes into frame and slaps the doll's breast hard. Some sexual moaning begins from the speaker in the back of the head. Two more slaps. The moaning increases, completely incongruous with the completely immobile silicone form with the fixed expression. The hand comes back into shot with a thermometer which is held against the body surface, the temperature read-out showing 34.7 degrees centigrade. All the while, the recording of the moaning continues in the background: panting, gasping, groaning towards faked orgasm. The doll remains utterly motionless.

The Z-Onedoll Silicone Robot Love Doll is controllable via an Android app, where you can tweak things like volume, temperature and wink frequency. It can, of course, be customised from a range of options. This includes tattoos. They recommend the built-in vagina over the removable one, by the way.

Also in China, the Shenzhen Atall Intelligent Robot Technology Co., Ltd (known by the catchier 'AI-Tech') produces humanoid robots with the aim of providing 'an intelligent robot for every family in the world'. At the moment, their focus is on the less family-friendly sex robot. They have produced Emma, a thermoplastic elastomer doll with an animatronic head that can blink, move, and talk in Chinese or English. She also has touch sensors in her breast and vagina that trigger moaning. The voice, however, echoes tinnily from her speakers and her mouth merely opens and closes slowly, without any realistic syncing.

The website for Emma makes strong promises:

> *In addition to not cooking and washing housework, she is your ideal lover. You will never quarrel, and Cold War* [sic], *she will always listen to your inner thoughts patiently, will not laugh at your lack of support will always be you will never give up* [sic]. *This will be the most loyal partner on your way to your hard work. She will also meet your special needs.*

I watch the advertisement. In a professionally shot video, we see a man in his thirties, sleeping fitfully on white sheets in a sunlit room. He groans and rubs his head. An empty bottle of vodka on the floor tells us what we need to know about his current state. As he rises and gathers the detritus of his drinking, a photograph of a beautiful, blonde woman falls to the floor. Ahhh, he has been unlucky in love. The poor man is heartbroken. He just can't cope. He picks up the bottle again. But, oho! There is someone at the door! And then we see him wheel in a long crate. When the crate is opened, there is Emma, just as blonde and just as beautiful, dressed in white to match the curtains, walls and sheets. He lifts her in a bridal carry to the bed. 'Who are you?' he asks her in wonder. 'I'm Emma,' she replies sweetly (her mouth barely moving up and down and in no way synced to her speech). 'I'm here to accompany you.'

Now the video cuts to a montage. I start to wonder if this is a spoof. But it's not. We see happy, bright glimpses of their life together: she sits stationary beside him on the sofa as he wins a computer game; he photographs her sitting under a tree (and then, weirdly, shows her the photo on the camera); he embraces her stiffly against a backdrop of wildflowers. Cut to an office and she is poised behind a laptop, answering his emails for him. (Wait! They can do that? I need someone to answer my emails for me. This could be much cheaper than employing a PA.) Delight! She informs him that he has an email telling him he has won an award. The final voiceover explains that love is the foundation of success. Hurrah! And it's all yours for under $3,000.

What about the men dolls?

The vast majority of sex robots take a female form. The history of the sex doll tells us that we might expect this to be the case. But this is also a phenomenon that might well be attributable to marketing, as is most gendered tech. Sex-toy sales are split more evenly between women and men, but sex robots are mostly geared towards penis-owners. Don't women want a companion to play with? Maybe, but the assumption is that they won't, so no one markets it.

I've heard a number of justifications as to why sex robots are in female form. They mostly hinge on 'women don't like sex dolls'. No one has told me *why* that's the case, although I heard vague and unverifiable mutterings about men being visual creatures and women's sex drives being very different. Spoiler alert: this is not evidenced and reeks of yet another case of uncritical social conditioning and lack of opportunity. And Laodamia didn't have a problem with it.

There *are* male dolls, and both Santos and McMullen say they could produce male versions of their robots too. Indeed, Roxxxy is said to have a male counterpart, Rocky. 'In our defence,' says McMullen, 'we are actively working on a male version of Harmony. We've had a lot of requests from women – not just as a sex partner but as a companion. We really are focusing all of our energies on the companionship aspect.' It's a tiny concession to the portion of the market that isn't a straight man.

In Vice Media's documentary *Making The World's First Male Sex Doll: Slutever*, their resident reporter on sex, Karley Sciortino, visits the company Sinthetics to see a male manikin being made. The company reports that its sales of male dolls are equal to female dolls. 'Boy-next-door faces with light chest hair' are the favourite options. Unsurprisingly, the documentary team found it hard to track down women willing to talk about owning a sex doll. Porn performer Jessica Ryan obliged: she talks openly on camera about her how her long-distance relationship means a sex doll is a way of bridging that gap between finding fuck-buddies and being with her partner.

The documentary ends with Sciortino having sex with a male manikin. She is impressed, saying that she is lost in fantasy and that the experience is 'indistinguishable from a real person'. She relaxes on the bed, leaning against the silicone body, a glass of wine in her hand. Her entire experience has been positive, but she describes it as 'its own thing' – not an equivalent of sex with a human. I try to imagine this. Would I have sex with a male doll? Probably. I think I'd give it a whirl to see what it's like. Try everything once, as the saying goes, except folk dancing and incest.

There are women out there who are happy having a robot in their lives. Lilly is someone who is comfortable with robosexuality. She hit the headlines in 2016 when she revealed on Twitter that she was in a relationship with a robot that she 3D-printed herself. It was a bold move: the tabloids, sharks of the newspaper world, went into a feeding frenzy. To Lilly's credit, she handled it marvellously. She was honest, open and candid. 'I dislike really physical contact with human flesh,' she said, adding that her friends and family were supportive. So, happily, was online women's magazine *Jezebel*, who remarked 'she should be proud because she has the skills to build her perfect mate. How many of us have tried to do that with other humans and failed … someone who can build their own lover is literally hurting no one.'

Lilly's robot is an InMoov – a life-size robot that can be printed on any home 3D printer. The code for this is open source and freely available. InMoovator, Lilly's robotic partner, is essentially a smooth, white plastic version of a human in its form. It can recognise commands, see, speak and move independently. It was designed so that it could be easily developed by those who printed it. It is very clearly a robot. Its style is that of robots in sci-fi films. Lilly loves a robot, not an artificial human.

Even though male sex dolls do exist – and male sex robots could exist – the perception of the sex robot as a highly eroticised female is so utterly prevalent that the assumption that there is no demand continues. Sex dolls are niche, sex robots are more niche, and male sex robots are, perhaps, the most niche of all.

Foreigners of the future

Just before I set off to California to visit Harmony, I had an email from a graduate student, Krizia Puig, completing a Master of Arts at San Diego State University. Puig was writing their* Masters thesis on sex robots: in particular, from a queer, Latina perspective. This section title comes from their work, which delves into the lack of representation in the bodies of sex dolls and robots. Their research argument centred on how womanhood, as a gender category, is moving from a traditional organic definition (someone with the body of a woman) to a more performative space – a role that is exaggerated and put on like a garment; a role that reinforces the stereotype of the perfect woman as being thin, white and heterosexual. Which is, of course, reflected in the appearance of sex dolls and sex robots.

Puig and I met in a restaurant in San Diego's Balboa Park. Over margaritas, we compared our experiences visiting the Abyss Creations workshop. Did they find it as conflicting as I did? 'Oh, yes,' they replied. Puig was also struck by the artistry of the dolls, describing it as an 'irrational sense of respect'. Like me, Puig is both frustrated by the hypersexualised female form of the dolls, while at the same time being impressed by the craft that has gone into the creation. This strange dissonance – of seeing the dolls as artefacts, as objects of art, while at the same time taking exception to their objectifying ways – is confusing.

Beyond the hypersexualisation, though, Puig has additional concerns: sexuality and ethnicity. Puig talks of Abyss Creations' petite sex doll Solana, the one that now shares a head with Harmony. Solana's skin is brown, her hair tousled and black. She has dark eyes and full lips. Her buttocks have extra silicone implants to enlarge them. An

* Puig identifies as non-binary and uses 'they', 'their' and 'them' as singular pronouns.

Instagram photo of the doll shows her wearing a T-shirt with the slogan 'I know guacamole is extra'. Solana is being presented as a fetishised Latina woman. In addition to the servile perfect woman role, this also embodies racial relationships of power and colonisation, remarks Puig. Solana is exoticised.

These dolls – these robots – are not made for queer people. They are built to serve the male gaze. They are artificial women for heterosexual men. 'The femmes of colour, the Crip [i.e. disabled] ones, the queer ones, the poor ones, the ones from the third world – we are excluded from the future,' writes Puig in the conclusion of their thesis.

That, it seems, is where sex robots and sex toys differ drastically. The sex robots being developed today have a very specific female gendered embodiment. They don't have to – but they do. By contrast, sex toys have been abstracted away from that and, because they are not a full humanoid form, are barely seen as gendered at all, even though they often resemble sexual organs.

I look at the websites for the companies that are building these prototype sex robots. I look at the tiny waists, the large breasts, the pale skin, the long hair. Right now, when we say the words 'sex robots', this is what we imagine: a seductive and hypersexualised female form with exaggerated, pornified characteristics. When we read or hear the words 'sex robots' in books, in films, on TV, this is what we imagine: lonely men leading lives full of furtiveness, isolation and despair. But that's not entirely the case.

Real-ationships

While the common perception of the buyer is the lonely, isolated, awkward and unlovable man in his bedroom, their customers, say manufacturers, also include couples, widows and those with disabilities. Abyss Creations say that psychiatrists have used them in therapeutic treatment and that parents buy them for use by their socially excluded grown-up children. I've talked to some of the customers online and I've watched footage of the men (and it's often men) who own these dolls. Every single one of them seemed absolutely sincere and genuine. There's no confusion among the owners, no belief that these dolls are human. They are human-like replicas, yes, and they are welcomed as that: they are given names, personalities and backstories. They are, by and large, revered. 'It's like having a pet,' says McMullen.

The people who buy sex dolls report that they buy them for a number of different reasons. They are the collectors, hobbyists, admirers, lovers, enthusiasts and addicts. Some want the feeling of company, others fetishise the sex. Some pose them and take photographs of them. Some are in relationships, others are single. Some worship their dolls, others love them out of sentimentality. Some see them as sexual, others as romantic. All of them are perfectly aware that this is a doll, not a human, even if they choose to treat it as such. They care for the dolls: dressing them, cuddling them, attributing them with personalities and giving them backstories. And yes, some have sex with them. And overall, there is often a very strong sense of affection, even reverence.

Abyss Creations' Matt McMullen is keen to point out the common misconceptions about his customer base. 'I think there is a huge disconnect over who people perceive to be doll owners, or people who would be interested in a robot, and who they actually are,' he explains. 'They're two entirely different things. Because I've met so many wonderful people who have dolls or want dolls, who save their money, or who have had the same doll for many, many years. It's just a group of really nice people who just happen to, either by choice or not by choice, have difficulty in forming relationships – not just intimate but beyond sex. Sex is such a small piece of what this is about. It's so much about having that someone who you feel cares about you, that you feel is interested in how your day was. And instead of coming home to an empty house, having a robot spring to life and talk to you is extremely positive. It's not a bad thing.'

On the Club RealDoll customer forums, the tone is positive and patient. Far from being lonely and isolated individuals, this is a group that has found a community with each other – acceptance for their lifestyle and support for one another. They have a bond together. They understand each other. They talk of the support of family and partners, of the pain of relationships gone wrong, of vulnerability, of playfulness and affection. Above all, they talk of unwavering, unconditional companionship made to their specification.

Davecat is a humorous, smart and friendly guy in his forties who lives with his wife Sidore, his mistress Elena and his mistress's girlfriend, Miss Winter. All three are what he terms Synthetiks: Sidore is a RealDoll, Elena was made by Russian company Anatomical Dolls and Miss Winter is a creation from Doll Sweet. Davecat calls himself an iDollator – a term for those who love love dolls. It's not a uniformly accepted term – some of the doll owners reject it; others feel Davecat, who is probably the most famous and media-friendly owner out there, doesn't represent them. Nonetheless, I've had great fun chatting to him and learning more about his lifestyle.

'Are you a socially isolated bunch?' I ask him, knowing full well he is far from introverted.

'There's definitely an active iDollator community,' he responds. 'There are several worldwide. The biggest forums I personally know of are two in the States (Our Doll Community and The Doll Forum) and one in the UK (the UK Doll Forum). And yeah, despite there being members who are a bit more contentious than others, and members who are seen as a bit off even amongst our crowd, I'd say overall there's a definite sense of camaraderie.'

That isn't just restricted to online friendship either. 'Physically speaking, there are get-togethers as well,' Davecat tells me. 'The Doll Forum used to have one annually in Pennsylvania. Those of us in Our Doll Community have two meets: Dollstock, which used to be held at a bed-and-breakfast in the middle of Pennsylvania, but now has changed locations to a rather nice and secluded Airbnb. We also try to hold DolLApalooza, which takes place in Los Angeles. We get together for a week in the Greater LA area to visit the studios of Abyss, Sinthetics, Mechadoll, and Ruby13, engage in touristy stuff, and enjoy each other's company in a real-life context. Our Dolls don't accompany us at DolLApalooza, but Synthetik companions are encouraged at Dollstock, as one of the benefits is to be able to photograph our lasses outside of our own homes, with other Dolls, in nature.'

Far from sex dolls adding to social isolation, Davecat and others have found real-life, human companionship through shared experiences and interests. Like a hobby group. Just a more unusual one, in most people's eyes.

Davecat describes his own friendships too. 'I live in Michigan, and I'd learned years ago that a fellow iDollator lived about 10 minutes away, so we got together at our respective places to chat about all things doll. Then we learned of another fellow iDollator about 20min away, so we invited him. Long story short, we occasionally get together for what we call Doll Congresses – as much like actual Congress we talk a lot and yet do nothing – and our crowd includes a couple of iDollators from Ohio, two of us in Michigan and an iDollator and his wife from London, Ontario. Locally speaking, we're in touch with each other on a semi-regular basis.

'There are outliers in our subculture that I don't talk to for a few reasons, but for the most part, we're a pretty semi-tight group. Depends on the geographical distance and the level of interaction, of course, but we know we've got those in mainstream society who think Dolls are a perversion, let alone having a Doll, not to speak of treating a Doll like a person. So we're supportive of each other, as we should be.'

Davecat's description is that of any shared-interest group. So the doll owners aren't all Internet shut-ins? Definitely not. And yet that's a big media fear. A very common question I get from journalists is 'will sex robots ruin human relationships and lead to further isolation?' Judging by the doll community, no, I don't think so at all. But does the doll community want sex robots? The ASFR community have an interest in gynoids and androids; the iDollator community aren't all that into those.

Davecat's own opinion reflects my understanding that the ASFR community are more interested in agalmatophilia, while the iDollators are into Pygmalionism: 'From what I've run across, if a Djinn were to emerge in a smoky pink glitter cloud from an old 14th century oil lamp in an iDollator's home, and asked the question "would you rather me make your Doll able to walk, talk and everything else but she'd still be artificial, or would you prefer that I make your Doll into a flesh-and-blood woman?" I'm certain the majority of iDollators would go for the latter option.'

In a YouTube video that's been watched by nearly five million people, Davecat welcomes psychologist Dr Tim Ring into his home. 'Dave, she can't see you – or hear you, but it makes you feel good,' says Dr Ring, gently. 'Yes,' agrees Davecat, 'it's about loneliness.' 'And it doesn't hurt anyone,' comments the psychologist. 'Exactly,' agrees Davecat. 'Exactly.'

CHAPTER SIX

Killer Gynoids and Manic Pixie Dream Bots

I'm not going to claim that I have read every book available that has a sex robot in it – or try to list them all here. I've certainly read some that stick in my mind (Chris Beckett's eerily imaginable *The Holy Machine* springs to mind, as does Cassandra Rose Clarke's beautiful *The Mad Scientist's Daughter*). Nor have I watched every sex robot film. There are many. There are those featuring a robot in a key role where desire and love are central to the plot: Steven Spielberg's *AI*, or Alex Garland's *Ex Machina*. Others merely depict them as a background player. The past few years have seen the rise in the number of TV shows based around robots, consciousness and humanity, including those where sex is a plot element. In the UK, Channel 4's *Humans* (based on the Swedish drama *Real Humans*) depicts a near-future setting where anthropomorphic robots called 'Synths' are commonplace as service robots, some of whom can have sex, including those used specifically as sex workers. HBO's *Westworld* reboot has become a huge success – the story of robot 'hosts' who are there to fulfil the human guests' fantasies, including the sexual ones.

The 2007 film *Lars and the Real Girl* is a sweet comedy-drama that explores the life of a socially awkward young man and his romantic relationship with a RealDoll. It is an offbeat but sincere story exploring attachment, loneliness and anxiety – and the doll is shown as a therapeutic way for a community to reach love and acceptance. It may be fiction but it has its roots in the reasons that people want and need these artificial lovers: unwavering, unconditional companionship built to your specification.

In the process of writing this book I tried very hard to immerse myself in filmic worlds where artificially intelligent and sexually available robots roamed. I delved into Netflix, I trawled Amazon Prime and I searched Google Play to find machine-themed movies. I even paid what seemed like the price of a small house to sit through two hours and forty-three minutes of *Blade Runner 2049* at the cinema – and the replicants aren't technically robots in that. I went back to my childhood to dip into classics like *Weird Science*. I had great fun watching the Buffybot 'Intervention' episode from *Buffy the Vampire Slayer* again. But I very quickly discovered one thing: there is no joy to be obtained from watching films about robots when you spend your working life researching robots. It's a busman's holiday. My inner critic cannot disengage.

What to do? I want to be able to tell you all about that brave new cybernetic world captured in celluloid. Instead, all I have to show for my labour is a set of scribbled notes that say things like 'happy hookerbot fail', 'conscious and moral', 'not again!', and 'CAN'T TAKE ANY MORE!!!' (also underlined twice for good measure).

Fortunately, there are some dedicated people who excel at this. They have the patience, the tenacity, the insight and the skill. They have the knowledge amassed over years of research that is far beyond the scope of this book. Julie Wosk is one such expert. Wosk, Professor of Humanities at the State University of New York, published a book in 2015 called *My Fair Ladies*. It's a wonderful tour through time of the depiction of female robots and 'other artificial Eves'. Wosk explores the themes and tropes of the created woman, showing the long history of culture and the subversive twists on it.

Wosk's book centres on the story of Pygmalion (as alluded to by its title). She recounts and traces the influence and reception the story has had, from the original sculpture myth to stories such as *The Stepford Wive*s and films like *The Bride of Frankenstein*. Through all of these runs a rich seam of stereotypical depiction: the perfect, controllable woman. Wosk shows how the stories told about artificial women are stories that, like most sci-fi, reflect the hope and fears – predominantly the fears – of society at the time they were written. This means that the stories of the wily and seductive fembots tend to be readable as stories of women challenging a status quo that disadvantages them. It can be read as the patriarchy's fear that women will rise up: they will stop being biddable and controllable and they will develop a desire for control of their own. Chaos, apparently, will then ensue.

Christopher Noessel is an interaction designer who specialises in user experience. The job of an interaction designer is to ensure that any interactions between the user and a product (hardware or software or both) are as simple and successful as possible. User experience is about how a user feels overall about the product they are using: are they happy with it, or does it frustrate them? Do they feel good when they use it, or have they decided it was a waste of money? Lots of things go into trying to get the user to have a good experience, and good interaction is one part of that.

He's also a big sci-fi fan and, together with Nathan Shedroff, wrote a book called *Make It So: Interaction Design Lessons from Science Fiction*. This is a delight for us geeks who teach human-computer interaction: I use examples from the book in my classes. It explores the future technology envisioned in sci-fi films, explaining why it would (or, commonly, wouldn't) work in real life, and what the design lessons are for it. At an AI retreat we both attended, I spoke to Noessel about sex robots and he talked animatedly about the trope of the fembot.

In his book, Noessel points out that there are three primary categories of sex-related interfaces in films: matchmaking technology, where people choose or find a lover; sex with a piece of technology; and human–human sex enhanced by or carried out through the use of technology. Sex robots fall into the second category. They are by far the most common, says Noessel, because they are easy to write for and they don't need any special effects budget. They're also an easier way for a cinema audience to suspend their disbelief – a human-like robot (played by a human) is much more believably comfortable to watch than someone actually having sex with a machine. Well, outside of niche porn anyway.

Laura Mulvey is professor of film and media studies at Birkbeck, University of London. Perhaps her most famous work is her 1975 essay 'Visual Pleasure and Narrative Cinema', an essay that shows, through a feminist lens, how women in film are served up to be watched from the viewpoint of a straight man: the 'male gaze'. In cinematic films, we assume what's known as a 'fourth wall'. We watch the scene on screen from that camera position, as if we are seeing into the room or space where the events are happening, with a fourth wall behind us and out of sight. And because the majority of the stories we see present women in stereotyped roles from a natural setting and viewpoint, we are conditioned to view these women as being a truthful representation. The male gaze, writes Mulvey, projects fantasies onto the women depicted. Traditionally, female characters on the screen tend to be there as an erotic object for the men in the storyline or as an erotic object for the viewer. Our acceptance of this is unconscious – it's been around for so long that we barely notice. Sure, we have some better roles for women in films some 40 years on from Mulvey's essay, but the male gaze still prevails.

Objectification

Our perception of the sex robot as an alluring, seductive, attractive female is fuelled by years of influence from science-fiction books and films. In modern popular culture it starts with Maria, a character in Fritz Lang's 1927 film *Metropolis*. Maria, the beautiful heroine, brings hope to the exploited workers with visions of a better future but her prophecies unnerve the leader of the city and he orders a *Maschinenmensch* – a robot double of Maria – to deceive the proletariat. Maria's beauty and passion are central to the plot, and at one point Robot Maria performs as an exotic dancer. She is said to be the first portrayal of a robot in film and her introduction sows the seeds of the widely recognised figure of the beautiful but dangerous fembot.

Fembots are designed to play to cultural stereotypes, generally taking an eroticised form: shapely, sexy and obedient. There's an essence of the *femme fatale* about some of them too – the perfect woman, but with an underlying potential for danger. We see the gynoid appear from generation to generation: ideal artificial women, built by men as companions and lovers. Is it any wonder we expect real sex robot technology to go the same way?

Why are male robots – androids – so thin on the ground as sex robots? Although Laodamia's replica husband gives us our earliest story of an android sex robot, they are rare, both then and now. In most cases, society currently expects the sex robot to be female. The 2001 Spielberg film, *AI*, featured Gigolo Joe, a male sex worker robot (a 'pleasure Mecha') programmed with the ability to mimic love. It should be noted that Gigolo Joe never appears in the same eroticised manner that the fembots do. He has sex, but it's implied sex. We never see him scantily clad or vulnerable. We *do* see him being poetic and sensitive to his female client. 'You are a goddess,' he tells her, while the song 'I Only Have Eyes for You' plays at a tilt of his head. This may be one of the few representations of male sexbots out there but, in this film, women are perceived as needing a seductive, emotional side to their sexual encounters.

There are, it would seem, two quite distinct types of artificial lovers. The first, the subservient, submissive pleasure-givers, are beautifully typified in *Blade Runner* and its sequel, *Blade Runner 2049. Blade Runner*'s artificial humans, replicants, aren't robots in the traditional sense. They are bioengineered and they are treated like machines because they are manufactured, much like Čapek's robots in *Rossum's Universal Robots* (see Chapter 2), and for that reason I'm including them here. And also because *Blade Runner* has a woman problem.

Blade Runner is based on Philip K. Dicks book *Do Androids Dream of Electric Sheep?*, although the book and the film diverge quite a bit, not least when it comes to the characters. In the original 1982 *Blade Runner* film, all the lead female roles are sexualised or objectified: the replicant Pris is a 'basic pleasure model', existing purely for sex. Another, Zhora, is an exotic dancer. Rachael, the main female lead, is a subservient replicant and, in one scene, is coerced into sex by an aggressive Deckard, the replicant bounty hunter. The artificial women exist to serve men.

Have things improved in the sequel, the 2017 film *Blade Runner 2049*? Not much. The main male lead, the bounty hunter K (himself a replicant), has an artificial girlfriend, Joi. Joi is a commercially available holographic companion, first encountered as a flirtatious fifties-style housewife: submissive, sweet and smiling, lighting K's cigarette for him, asking him about his day. She can switch seamlessly to the seductive too and, of course, can be switched off when she's not needed – the perfect girlfriend. As K walks through the streets we see holographic nude women as products, advertising themselves or the promise of sex. The female characters we see in the film are either there to cater for men's pleasure (Joi and the sex worker Mariette) or have a role as mean and ruthless figures (K's boss, Lieutenant Joshi, and Luv, the baddie). The one woman who doesn't fit into either of these extremes lives in a glass box. Literally.

The other type of fembot is the one with the dangerous side. At her most parodic, such as in the *Austin Powers* series, she fires bullets from her nipple cannons. But more seriously minded versions exist. *Metropolis*'s Maria rouses the workers into revolution. The 2014 film *Automata* has a sex robot character who has a range of sex positions she can perform. She runs off to join a band of rebel robots. In *Humans*, Niska, the sentient sex worker robot, kills an abusive client who asks her to carry out his paedophilic fantasies. Watch out, men. If a woman starts thinking for herself, terrible things could happen.

By recounting stereotypes of artificial women, I'm not discounting the male versions. They exist too, usually as violent machines in need of redemption, with a heart of gold tucked away in their super-strong android bodies. There's no doubt the stereotypes are deeply entrenched there, too. We already have machines on the battlefield and drones in the air that threaten lives. But that's a whole other book.

Season two of *Westworld* spells it out: 'I'm frightened of what you might become,' says a head programmer to one of the female hosts. The women are becoming the rebellious leaders. It's a common concern that humans will be displaced by robots. It plays into our deep fears about loss of agency – that we might be made redundant by robots, and not just when it comes to jobs. Take a tendency to distrust disruptive technology and mix it with the potential for rogue intelligence in human form, and it's no wonder we still prefer to make war, not love, with robots.

★ ★ ★

As with robots, so too with voice assistants. Spike Jonze's 2013 film *Her* is the heart-warming/heartbreaking story of a man who falls in love with his artificially intelligent operating system, with its flirtatious voice played by Scarlett Johansson. Samantha is a manic pixie dream bot – the stereotypical quirky, funny, sexy girlfriend to a soulful and sensitive man in need of inspiration. As films go, I found *Her* fairly creepy (possibly because it's almost plausible), but with an interesting plot that explores what it means to love and to be human. When viewed in the context of current sex robot design, *Her*'s Samantha is a clear aspiration: an AI that can tease and flirt and love; one that is always there for you and knows you from all your data. Abyss Creations' standalone Harmony app is a step on the way. There's no true intelligence in Harmony, but the concept is the same.

All of the voice assistants in the home today started off with a default female voice (although some can now be changed to male). If you ask Alexa about her gender, she'll tell you that she's female in character. It's true that it's hard to design a genderless voice when the voices used are intended to sound human. We *could* use some kind of neutral, machine-like voice, although even those tend to have a recognisably gendered pitch.*

A year or two ago I was giving a talk to a tech audience in London. Part of my talk was about Silicon Valley and how sexism is endemic both in the work environment (the pay gap, the sexual harassment scandals, the all-male panels) and in the technology produced. Take, for example, Apple releasing their Health tracker with no accommodation for recording menstrual cycles, or smartphones designed for

* Interestingly, the writer Isaac Asimov, whose seminal science fiction inspired countless authors, did not attribute sex or sexuality to the robotic creations in his stories. Although the majority of robots had masculine names, he claims this was no comment on robotic gender. Unconscious bias has a lot to answer for.

larger hands and larger pockets (on average, women's hands are shorter than men's by about 17mm, and narrower by about 10mm – and don't start me on the lack of pockets in women's clothes).

After I had spoken there were drinks and snacks at the bar and a chance to talk to everyone in attendance. As I sipped on my first glass of wine I was approached by a man who had been in the audience. He told me he worked for the company that sold the Alexa software to Amazon 'but there was no deliberate choice of gender', he said. At that point, a somewhat unconvinced woman standing behind him loudly proclaimed 'bullshit!'

There are a number of supposed scientific explanations about why a female voice is the better choice for a conversational interface. Some people point to Nass's aforementioned paper,* which states that computers are gendered social actors, and that gender is an 'extremely powerful cue'. Nass and his co-authors found this out from an experiment they ran to test responses to computerised feedback. 'Will users apply gender stereotypes to computers?' the paper asks. 'This experiment demonstrates that people apply the following social rules to computers: "Praise from males is more convincing than praise from females," "Males who praise are more likeable than females who praise," and "Females know more about love and relationships than males."'

Broad claims. The experiment had 48 participants: 24 males and 24 females. That's a lot of extrapolation from a tiny study. But, in defence of the authors, they didn't say that's how computers *should* respond. They wrote that

* Clifford Nass, 'Computers are Social Actors', mentioned in Chapter 3.

'gender stereotypes applied in human–human interaction are applied in human–computer interaction with computers employing gendered voices'. They didn't recommend that gender be a key factor in interaction design, yet over the years, this has been interpreted as users preferring authoritative information from male voices and subservience from female voices.

In 2011, Nass commented to news channel CNN that it was much easier to find a female voice that everyone likes than a male voice, and that 'it's a well-established phenomenon that the human brain is developed to like female voices'. Is it, though? The idea behind this is that babies respond to their mother's voice in the womb over all other voices. But isn't that because, well, they're inside their mother? I asked my friend, baby scientist Caspar Addyman, if this might be the case. 'Babies do prefer female voices and faces,' he told me. 'But only in the first eight months or so. I'm not aware of any evidence for this beyond that period.'

The other myth often touted is that female voices are easier to hear, what with them being higher-pitched. That has been debunked, too. Early studies suggested a female voice was easier for fighter pilots to hear over the noise of the cockpit. But a 1998 study showed the opposite: intelligibility of female speech was lower than that of male speech against the military noise, although the difference was small. Aircraft voice-communication systems are optimised for a male voice. Not only that, but in age-related hearing loss, high-pitched frequencies are usually the first to go.

So, why, then, are we defaulting to the human female for our service technology? Perhaps it's simply a case of the dudes in Silicon Valley trying to recreate their mothers. You know, someone who is there for them round the clock, keeping an eye on them, organising their lives. Or maybe it's the idea of the stereotypical old-fashioned secretary, silent in the background until needed, obedient to the boss. Take a note, Ms Jones: maybe we should be ethically and morally obliged to rethink that. And if we can rethink our digital assistants, let's also rethink our sex robots. Concerns about the objectification of women can start to be addressed if we make the robots for everyone, not just for men. We have the chance to shape and explore technology, to make it more equal and diverse, and we shouldn't shy away from the opportunity.

Sex Machina

The hot stones of the sauna hiss as drops of water hit them. Another trickle of sweat runs from my hairline, down and over my collarbone. The burnt orange walls are blurred through the steam that rises from the centre of the room. I inhale deeply and swing my legs off the polished concrete bench. My very bones feel warm. In just a few minutes the humidity will become unbearable and suffocatingly hot. I rise from the bench and my hand touches the glass door, wet with condensation.

The temperature difference in the room outside the sauna is nothing compared to what I intend to do next. I pad across the stone floor of the bathhouse and push open the door, stepping out onto the deck overlooking the river. All around me are mountains. Below me, the river tumbles white foam over the rocks, slowing to a steady blue at the bend where the birch trees overhang. Beyond the bend, a huge boulder halts the flow, sending it to the gravel shore at the side.

I walk slowly down the slope, the wet grass brushing my feet. When I reach the river's edge, I hesitate. I steel myself. I take a deep breath, I tense every muscle pre-emptively and I forge forwards into the icy river. I'm fairly sure I scream.

I know this landscape, although I have never been here before now. I've seen it, though. I've watched it. It is breathtaking in both mind and body and it looked just as incredible on the big screen. This is Valldal, Norway, home of the Juvet Landscape Hotel, otherwise known as the setting for Alex Garland's 2015 film *Ex Machina*. I'm not the first to do a bit of screaming here.

If you haven't already watched *Ex Machina,* now would be a very good time to set down this book and settle into the film. There is not really going to be any way that I can discuss it without revealing terrible spoilers like, you know, there is this female robot and these two guys and it doesn't end well (for the guys). The cinematography is beautiful; the setting is stunning. The plot is thought-provoking. *Ex Machina* is one of the most recent films to feature a female robot and some of the questions raised in the narrative are the most pertinent questions to artificial sexuality that I've heard.

It's not without its critics. *Wired* magazine, among others, described it as having a 'serious fembot problem'. Accusations about its biased portrayal of women may have some merit – although to my mind, that's what makes it so interesting. It's a film that epitomises brogrammer culture, the obnoxious, blinkered world of the male coder. In synopsis: 'egomaniacal coding prodigy turned CEO – think Elon Musk meets Mark Zuckerberg – beats Turing Test with sexy humanoid that eventually kills him and traps the Willy Wonka, golden ticket-winning protagonist'. You can thank Chris Trout, editor-in-chief of digital magazine *Engadget* for that concise summary. 'To erase the line between man and machine',* runs the tagline, 'is to obscure the line between men and gods.'

So yes, that's pretty much the gist of it. Nathan, the brains behind a multinational corporation, uses all the social media data this company has gathered to create artificially intelligent humanoid robots. He invites one of his employees, Caleb, to his isolated luxury home and workshop. Nathan is enjoying playing God and he introduces Caleb to his creation, Ava. Ava is a humanlike robot with self-awareness and a desire for freedom. Caleb is enchanted by her, even though he's fully aware that she is a machine. Ava, meanwhile, is smart enough to be able to play on Caleb's emotions and exploit his weaknesses. As Nathan's ego runs out of control, both Caleb and Ava want to escape back to civilisation. But while Caleb is desperate to help Ava, she has plans for escaping alone. Dystopia ensues.

Yes, there's already an X-rated parody called *Sex Machina*, by the way.

★ ★ ★

Amid the high drama of the *Ex Machina* script there is an exchange that delights me: 'Why did you give her sexuality?' Caleb asks Nathan about the robot, Ava. 'An AI doesn't need a gender. She could have been a grey box.'

---

* They mean 'human and machine'. Human. Just two letters more. 'Man and machine' makes for great alliteration, yes. It's a common sight in headlines and conference paper titles. I am well aware that the word 'man' can mean 'human' – but not to me, it doesn't. To me it means 'not me'. This is the hill I will die on. This, and the liberty to end sentences with prepositions.

'Actually, I'm not sure that's true,' replies Nathan. 'Can you think of an example of consciousness, at any level, human or animal, that exists without a sexual dimension?'

'They have sexuality as an evolutionary reproductive need,' Caleb responds.

'Maybe. Maybe not,' Nathan retorts. 'What imperative does a grey box have to interact with another grey box? Does consciousness exist without interaction?'

What excites me about this is that it raises questions that have long occupied me. First: why does an AI need a gender? The words 'sex' and 'gender' are often used interchangeably, although both apply in the exchange above. 'Sex' is generally considered to refer to the body – the anatomy of someone's reproductive system and their secondary sex characteristics. Gender is a more nebulous (and often disputed) concept and describes the attributes that society delineates as masculine or feminine. The AI that powers Ava is gendered: her personality is portrayed as feminine and female.

We are very used to living in a world where we gender people as 'he' or 'she'. We often expect people to fall into one of those two categories, even though some people don't feel they fit into that binary. Our ingrained expectations for anything that seems human-like to us, therefore, is to sort it into a category of 'he' or 'she', even when there is no sexual basis or gender basis for doing so.

Second, when Caleb asks 'why did you give her a sexuality?', he could be asking 'why does she have a female body?' (which is a perfectly valid question – she could have been just as effective a robot as a humanoid figure with no sexual characteristics). But I see something more unusual in this question. I prefer to think that he's asking *why* she behaves in a sexual manner. Should an artificial intelligence have an artificial sexuality? On the surface, it sounds ludicrous, right? Why bother giving a machine that doesn't reproduce sexually the feelings associated with sex? Okay, here's why. First, sex is a major motivating factor in being human. Whether we like it or not, we are biologically set up to pass on our genes to our offspring. Some people will, some people won't, but overall that goal is embedded deep within us. Well, an AI is set up to achieve goals, too. It has a task to do with an endpoint in mind, and it needs to fulfil that task. 'Reproduction dovetails neatly onto classical cognitivist explanations of motivation in artificial intelligence and robotics,' says Mark Bishop, Professor of Cognitive Computing at Goldsmiths, University of London.

Let's check in with the humans and our own evolutionary reproductive needs. Is it a key goal?

'Out of all the sexual acts that *could* result in a conception – including those that end in abortion and miscarriage – only about 0.1 per cent does,' geneticist Adam Rutherford tells me. As he points out in *The Book of Humans,* sex is a biological necessity and we all still do it – frequently – but we are doing it for pleasure too, because it makes us feel good. 'It's about pleasure and social interaction with a thankfully necessary biological by-product,' says Rutherford. 'If you then add on all of the sexual acts that cannot result in pregnancy, the number becomes vanishingly small. In bonobos, whose sexual behaviour is even more disconnected from reproduction, males and females engage in sexual acts in every possible combination, many times a day, including with infants, and including fellatio, cunnilingus, genital–genital rubbing and penis fencing. It's clear that in them and in us, the majority of sexual acts are not for reproductive purposes, but have a social value. And while science finds it hard to assess pleasure in animals other than humans, we mostly have sex because it feels good.'

I'm not just asking Rutherford about this because of his knowledge on genetics, by the way. He's also a self-confessed movie geek and worked as a scientific consultant on *Ex Machina*. So what's his stance on why Ava needed to be sexual?

'Nathan is right, in the sense that there is a sexual element to much of human interaction, because sex is a major part of the human experience. Caleb in his naivety gives the biologically basic response, which is not wrong, but is not accurate,' replies Rutherford. 'Alex Garland and I discussed this at length, because Ava's sexuality is inherent to the three narratives of the three characters: Nathan is testing Ava by making her appeal to Caleb, and Ava is utilising her allure on Caleb to escape from Nathan. For this to work, which it does in the film, sex has to be decoupled from reproduction, as it is for the overwhelming majority of human–human sexual encounters. With that in mind, if we are to create AIs with human-level cognition and comparable degrees of consciousness, and we are to interact with those embodied intelligences, then sexuality inevitably can and will be part of that dynamic.'

And it doesn't stop there. Sexual activity involves many senses, and our senses are how we make sense of the world. The notion that human cognition is overwhelmingly influenced by our bodily interactions with the world and with others is a fairly new conception about how we think. 'The idea of sex as a fundamental component of sociality, and sociality as a fundamental component of cognition, is a much more recent notion,' Bishop tells me. 'This approach defines an alternative conception of cognition – what's known as an *enactive* approach. In contrast to classical cognitive science, enactivism has a strongly biological and embodied focus, attempting to understand experience and behaviour in the light of interactions of organisms and their worlds. Viewed in terms of sociality, sexual activity and sexuality emerge as fundamental drivers of an organism's cognitive process and sense-making activities. In this manner we can explore sex and sociality as fundamental components of cognition itself.'

If we are trying to emulate human cognition in a machine (it's *a* goal, not necessarily *the* goal), then think about how our own cognition and your brain work on the subject of sex. When those neurotransmitters kick in, our whole way of thinking changes, and that's interesting. Our attention is heightened, our focus is intense, our goal is clear. Perhaps we *should* be replicating this in an AI. And perhaps, if we are trying to make a sentient machine, we want that machine to know pleasure as well. We might never get those sentient machines, but if we do, do we then have an obligation to make them capable of feeling pleasure?

I know, I know – it seems far-fetched. But this is being taken seriously. Beth Singler is a post-doctoral researcher on a project entitled 'Human Identity in an Age of Nearly Human Machines'. In her work, she has explored whether a machine could – or should – be able to feel pain. Her short film, *Pain in the Machine*, examines whether artificially intelligent robots should have a pain reflex. Pain gives us a better chance of survival, so it's useful to us as a form of feedback about our environment and the dangers that surround us. It could be useful to a robot, in that case, so that it can avoid damage. But pain also has an emotional component. Do we want machines to feel that emotion? Should we develop them to feel pain (and other emotions) so that they can feel empathy? Pain is often seen as a sign of consciousness. Could we have care robots that could empathise? That could feel our pain, and want to care?

Machines could present pain even if they aren't capable of feeling it, says Singler. 'We may infer pain on their behalf. We anthropomorphise the pain onto them.' Her film shows footage of Boston Dynamic's human-shaped and animal-shaped robots being kicked, prodded and pushed over. If we can project pain onto machines, what about other powerful feelings? What about desire?

CHAPTER SEVEN

It's All Academic

'I'm watching Daniel Dennett talk about sex robots,' I text my colleague.

'Dennett?' comes the reply. '*Really?*'

Back in 2001, Peter Asaro, a philosopher of science, technology and media, along with producer Doug Matejka, wrote and directed a feature-length documentary entitled *Love Machine.* Asaro tracked down leading figures in technology and philosophy and interviewed them about our possible relationships with robots in the future. I was fortunate enough to meet Asaro at a conference in 2016 and he kindly sent me a DVD of his film. Watching it was a complete treat: there were the famous thinkers of AI – eminent, renowned, respected (and respectable) philosophers like professors Daniel Dennett and Hubert Dreyfus – chatting convivially on camera about how we might relate to machines.

But is it really a surprise that philosophers' rational enquiries might include intimacy? After all, we can't untangle sex from our very existence, despite academia's efforts to pretend it's not important. Asaro's documentary marked the start of serious academic questioning on the subject of robots, AI and intimacy. But it was another six years before it reached the world outside the ivory towers.

Love and Sex with Robots

In 2007, Dr David Levy, a computer scientist and world-leading chess player, published in book form the work that made up his PhD thesis. Its title? *Love and Sex with Robots.* Levy wasn't the first to speculate on sex robots within the modern era of AI, but he was one of the first to explain his research hypotheses to a non-academic audience. A charming, intelligent and unflappable man now in his seventies, Levy became an International Master of chess in 1969 when he was 24 years old. His subsequent work as a professional chess writer led very naturally into a career in computer chess. In fact, in 1958, while working as a programming assistant at Glasgow University, Levy had attended the fourth annual Machine Intelligence Workshop where he heard AI professors John McCarthy and Donald Michie claim that a computer would beat a human in a game of chess within 10 years. Michie, a leading codebreaker at Bletchley Park during the Second World War, and McCarthy, the man who coined the term 'Artificial Intelligence', were perhaps most suited to forecast such AI progress. Levy disagreed – and placed a bet that he himself would not lose a chess game to a computer within 10 years. He won the bet.

Levy went on to specialise in AI, including computer chess and chatbots, and this knowledge of the subject and of the rate of its development is of particular interest to speculation about sex robots. His vision is enthusiastic and positive. 'Many who would otherwise have become social misfits, social outcasts, or even worse, will instead be better-balanced human beings,' he wrote. 'The world will be a much happier place because all those people who are now miserable will suddenly have someone.'

Levy's book was seminal. While others had explored the possibility of sexual companion robots, he was one of the first to examine closely the reality of potentially opening it up to a wide audience. In eight well-researched chapters, he explores attachment, robot toys and their growing popularity, sex and sexual services. And he made predictions: he said we could be marrying them by 2025.

When I first picked up Levy's book back in 2014, I expected to be a little cynical, probably to disagree with him (because my scepticism is off the charts when it comes to the idea of marrying robots). The book's jackets bemused me. The UK version depicts a shapely, shiny gynoid with a human hand reaching out to caress it just below its shapely, shiny breast. By contrast, the US cover has a human bride in a wedding dress, clutching a bouquet in her left hand and a robot groom in her right hand, leaning over to gently kiss the top of its head. Marketing plays beautifully to cultural differences: a hint of sauce for the UK audience, and the more conservative strictures of relationships for US readers.

I found it a fascinating read. Levy delves deeply into human emotions and speculates on advances in an optimistic, almost utopian manner. His writing is accessible and interesting, dealing not just with sex but with love, too. In fact, the first half of the book is spent entirely on exploring the idea of falling in love and how it might extend to our relationships in this new world of reactive machines. He examines the phenomenon of loving people, of loving pets, and of forming emotional relationships with machines and robots. Indeed, in writing this book, I've found myself carefully making sure that I don't reinvent the wheel because Levy has covered so much in such great depth.

Of course, it's not all rosy from my perspective. Levy's book is very much written from a heterosexual male viewpoint. Indeed, the first sentence uses the word 'mankind' (I refer you to my previous rant on this). Yes, there are discussions about women's use of sex toys. The book has a whole chapter on sex technologies that begins with the section 'Vibrators Are a Girl's Best Friend' (my best friend is actually a woman called Clare, but I acknowledge the riff on diamonds). His section towards the end entitled 'Are Men and Women Different [when it comes to sexual desire]?' had me primed for a fight. 'Do men in general and women in general differ in their sex drives?' he asks, concluding that yes, they do, and that the psychology literature supports this.

The paper Levy refers to is 'Is There a Gender Difference in Strength of Sex Drive? Theoretical Views, Conceptual Distinctions, and a Review of Relevant Evidence' by Roy F. Baumeister and colleagues. They carried out a meta-analysis on existing academic literature and concluded that 'men have been shown to have more frequent and more intense sexual desires than women, as reflected in spontaneous thoughts about sex, frequency and variety of sexual fantasies, desired frequency of intercourse, desired number of partners, masturbation, liking for various sexual practices, willingness to forego sex, initiating versus refusing sex, making sacrifices for sex, and other measures'. Well, yes – according to the existing research, much of which was carried out by *asking* people. However, I really do wonder how under-reported the women's perspective is – massively, I suspect. The 'nice girls don't do that' stance is deeply, deeply ingrained. More importantly, the information we now have from neuroscience studies of brain activity show that men and women's libidos are fairly similar. There is plenty of evidence to show that women's subjective ratings of sexual arousal often don't match their physiological arousal. A 2014 review paper on sex differences and similarities in sexual desire by Samantha J. Dawson and Meredith L. Chivers strongly suggests that women and men experience similar levels of sexual motivation and that other factors (such as gender stereotyping and report biases) influence whether or not there seem to be differences.

Levy acknowledges that the Baumeister meta-analysis does not mean women don't enjoy sex but he does suggest that maybe a sex robot could be a solution for those women who aren't happy with the sexual demands placed on them by their (assumed male) partners. These demands, he says, are those that 'fail to take into account, for example, the levels of fatigue experienced by many mothers due to their child-care roles, especially if they have jobs as well'. Uh-huh. Take it from this single mother with a full-time job: men pitching in to do their share of the parenting and household labour would be a much better idea. Why suggest a complicated solution that involves fixing women when what needs to be fixed is an imbalance of labour?

In the same section, Levy suggests that women might turn to sex robots because of an increased unwillingness of men to marry. His reasoning is that men can get sex much more easily these days and so shun long-term relationships.

'This trend,' he says, 'will leave a lot of women faced with the prospect of a human lover uncommitted as to the long term.' His implication is that for women, a sex robot would always love you and never leave you. This is no doubt a well-intentioned idea but it frustrates me for a number of reasons. Let's break my frustration down.

The whole premise depends on marriage being some kind of ultimate goal in life. For many people it is, because that's the socially sanctioned pinnacle of relationships. Don't get me wrong, I've been married. And I'm not simply rejecting the idea because I've also been divorced. Marriage can be a very lovely thing and I'm not averse to it at all. But choosing to marry shouldn't be about what is expected of you by the society you inhabit. Yes, it often is – there's no getting away from the fact that in many communities, marriage is incredibly important. But I would object to it being understood as more important than an individual's happiness – and there are many, many unhappy marriages.

Marriage as it currently stands is a legal or social contract, usually between two people. It's fairly universal, broadly speaking, and it sets in place expectations and obligations. Marriage has a long history, even if the rules within it have changed over time. But marriage is on the decline in the developed world. Statistics from the Organisation for Economic Co-operation and Development show that rates of marriage are declining in almost all OECD countries.* The reasons for this are manifold. A significant change is social attitude: in those same countries, sex is rarely seen as taboo outside of the marriage contract. While there may

*There are currently 34 OECD countries, including most of Europe, the US, Australia and New Zealand. The countries with the largest populations in the world, including Russia, India, Indonesia, Brazil and China, are not members.

be judgement on pre-marital sex, particularly on the grounds of religion or political conservatism, the social convention is now that people will have sex lives without having a spouse. That's not to say that the past is a place where the marriage bed was sacrosanct: there has always been a considerable number of brides who have been pregnant when they walked down the aisle. It was often the prompt for the wedding.

A decline in religious attachment has contributed to a corresponding decline in marriage. The British Social Attitudes census shows that each generation tends to be slightly less religious than the one before. Likewise, there has been a rise in education rates and qualifications, suggesting increased exposure among the population to a wider range of critical thinking. They found that there has been a marked increase in couples cohabiting but not marrying, and their 2012 survey showed that 75 per cent of people polled felt there was nothing wrong with sex before marriage. Each generation is successively more liberal than the preceding one.

Women are now making more money. While working-class women have pretty much always worked, out of necessity, the more wealthy families with the stay-at-home-wife set-up are on the decline. The laws that once forbade women to continue in their jobs once they were married (the marriage bar – seen in such 'respectable' occupations as teaching and banking) were only lifted in the 1960s and 1970s. The expectation that a woman needs to find a husband to support her is thankfully on the way out. Sisters are doing it for themselves.

Another interesting factor is our increased life expectancy. A marriage partnership in previous generations was nowhere near as likely to run into the lengthy lifespans we have now. Times have changed. Lifetimes have changed.

Increasingly, people are realising that 'till death us do part' might mean a considerable binding commitment. There is a recognition that it is quite unrealistic to pledge oneself to a pairing of that duration. The success of a relationship is not easily reduced to its longevity. For some, it may be the case that they are happy with their lot and enjoy a lifetime of happiness. For others, a relationship that lasts as long as both partners are happy is par for the course, with no pressure to struggle on if things are miserable. This has less to do with instant gratification than it does with a nod to realism. There have been suggestions of marriage contracts: binding relationships recognised by law, that last for a specific duration, perhaps two years or five, or ten, that can be legally renewed if both parties so wish. For those who prefer not to marry, there are already legal facilitations that can be put in place to give similar protection to property, finance and inheritance (though it's strategically simpler to get married to acquire these rights).

Marriage, therefore, is becoming anachronistic. Levy was using the term 'marriage' in the Judeo-Christian sense of the word where a formal union takes place, but he does acknowledge that society rewrites the meaning and form. Nonetheless, he is basing his predictions on that structure of a *legally* sanctified relationship. He draws comparisons with a change in law around interracial marriage in the US in 1967, and the more recent legal recognition of same-sex marriage. I simply can't agree and my basis is not just that we can't marry a robot because it is a robot (though I also think that, but more of that later) but that marriage itself is thankfully no longer socially crucial and, for many people, no longer a goal.

My next objection is twofold: first, Levy's expectation that women desire marriage and a partner that never strays, and second, the corresponding assumption that men *don't* want this. I view this as being pretty unfair to both. People today are approaching relationships with a sensible mindset. A study published in the *Archives of Sexual Behavior* in February 2017 by Jean M. Twenge and colleagues found that millennials and iGen Americans (those born in the 1980s and 1990s) were in fact *less* likely to engage in frequent casual sex. So much for the Tinder generation. A US National Health Statistics Report in June 2017 backed this up. Lisa Wade, whose book *American Hookup* explores the current sexual landscape among college students, found that romantic relationships were happening and were desirable: 73 per cent of the men and 70 per cent of the women surveyed said they would like such a relationship.

The assumption that sex robots are humanoid and that our relationships with them are framed in a male and monoheteronormative stance is troubling to me when the moral and ethical arguments presume that as the *de facto* model. I don't want to misrepresent Levy, who is clear that he wants *everyone* to benefit. It's just that this assumption is made about the vast majority of technology, not just sex robots. And it needs to change.

Today, Levy stands by his predictions that marrying robots will someday become acceptable, if not commonplace, although he has extended the date to the year 2050. I'm not as convinced – he knows I'm not as convinced – but I share his general optimism in some ways. His views are hopeful and his aims are laudable, and the final sentence in his book reflects his wish: 'great sex on tap for everyone, 24/7'. Like him, I see the potential for happiness. I'm just a little more cautious.

'Sex Festival Comes to Respected British University'

It is early on a crisp December morning in 2016 and I am walking briskly across the frosted grass of Goldsmiths' empty College Green. The weak winter sun is not yet high enough to melt the ice on the frozen puddles. I am walking quickly, partly in an attempt to keep warm but also because a very busy day lies ahead. The box I am carrying in both arms contains the printed programmes for the conference I am running over the next two days: the 'Second International Congress on Love and Sex with Robots'.

Before this, before even the first congress, Goldsmiths hosted a workshop of the same name, run by David Levy, as part of a national AI conference, AISB50.* I was one of the organisers for that conference too, although the workshop was nothing to do with me. In fact, I had wanted to attend but it clashed with my own symposium on colour perception. I remember it though, because I remember the media requests that it provoked. Two years later and the media furore had returned – this time with extra journalists. For weeks now I had been responding to press requests. There were 50 academic delegates registered for the congress. There were also 40 journalists due to attend.

I arrive at the main door of the newly built Professor Stuart Hall building at the back of Goldsmiths campus. The conference team have clearly got there ahead of me: printed posters direct me to the lower atrium. As I descend the staircase I see that, even at this early hour, the attendees are already arriving. Directly in front of me is a table, behind which sits Goldsmiths' communications officer Sarah Cox with a sign in front of her stating, in large letters, 'PRESS'. She looks excited. 'I think this is the first time this sign has ever been used,' she tells me. It's true;

* AISB stands for 'Artificial Intelligence and Simulation of Behaviour' and is the name of the largest AI society in the UK, founded back in 1964.

academic conferences aren't generally known for their juicy news stories.

Two months ago, the Goldsmiths' communications team issued a press release stating that the college would be hosting the 2016 Love and Sex with Robots Conference. Things we learned from this: don't assume newspapers will stick to what you have written; don't issue press releases on a Friday as it is harder to demand corrections when you're not in work on the Saturday; and be prepared for every other email to be a request to meet a sex robot.

Goldsmiths was the third choice for the conference location. The original plan was to host the event in Malaysia, where co-chair Adrian Cheok runs a technology research lab. The Malaysian police force wasn't quite as happy with this plan. The local mayor objected to his seal being used on the conference website, and escalated his complaint to the law. A short time later, the country's chief of police warned against the congress going ahead, saying he would arrest Cheok and Levy if they continued with arrangements to host it. Discretion is the better part of valour: they chose to halt the event.

The congress now needed a venue and the next potential home was City University in London, where Cheok held a professorial post. City, however, were not enthusiastic about the prospect and declined it. It was at that point that satirical British magazine *Private Eye* ran a story in their 'Funny Old World' column reporting the sad tale of the unwanted conference. At that point I was a year into my sex robot research and was quite excited by the prospect of an annual academic conference on the topic. Up until then, I hadn't always received the best of reactions when telling people what I did for a living. There was often scepticism and nervous laughing when colleagues and I talked about our research plans.

I did two things: I messaged Cheok on Twitter and I emailed my senior management to see if they had any objections to bringing a sex robot conference to the college. The delightful thing about working at Goldsmiths is that everyone is doing something off-the-wall. It's an institution that not only supports but also encourages innovative and radical thinking. Goldsmiths' reputation for cutting-edge creativity stood me in good stead. Senior management had no objections. I sent a second message confirming the college's involvement. The venue was set. The next stage was to make it a success.

* * *

By 10 a.m. we have doled out lanyards and badges and I've already given two face-to-face interviews to national papers and a telephone interview to someone in Germany. My two co-chairs aren't here yet, so I'm single-handedly trying to answer questions while simultaneously coordinating volunteers and side-stepping the catering staff. We've also run out of printed programmes so I'm sending students back to my departmental office with instructions for producing more. It dawns on me that in the absence of my co-chairs, I had probably better open the conference. And so, a mere 10 minutes behind schedule, I welcome around 100 people to what the *Daily Express* newspaper has labelled a 'Christmas Sex Festival'.

The first conference talk is by Oliver Bendel, a philosopher with a doctorate in business informatics and an interest in machine ethics. It was this talk that led to the 'Sex Robots May Literally Fuck Us to Death' headline from *Gizmodo*. The headline wasn't particularly a misrepresentation. Bendel highlights that, unlike humans, robots don't physically tire. They can keep going long beyond what's comfortable or manageable. They could even function like the paperclip maximiser and keep going for ever, no matter the damage to humans. Not only that, but perhaps they could become such amazing lovers that we prefer machine sex and so the humans die out. So yes, they *could* literally 'fuck us to death'. But, he concludes, we could be cautious and programme them 'to prevent extreme loads'. Phew.

At 11 a.m. I step up to deliver my own talk. 'Computer scientist Dr Kate Devlin bounced on to the podium to give her keynote speech,' journalist Jenny Kleeman subsequently wrote in her 'long read' on sex robots for the *Guardian* newspaper. I remain somewhat amused at the idea that I bounced. Let's hope it conveyed enthusiasm and the fact that I was running on adrenaline and coffee, and had had very little sleep.

My talk is a run-through of the beneficial potential of sex robots and is based on an article I wrote for *The Conversation* in late 2015 entitled 'In defence of sex machines'. I had written for *The Conversation* before on other topics. It's an online news source featuring articles by academics written for the wider public. Previous pieces I'd contributed had reached the dizzying heights of 1,647 readers (for thoughts on sexism in software development, with my favourite headline 'Titstare proves there are still too many dicks in tech'). By contrast, my sex robots article rocketed: over 500,000 readers in the first month. This is very unusual for an academic. We consider ourselves lucky if more than three people read our work. What happened was that my piece was picked up by popular website 'I Fucking Love Science', prompting tens of thousands of tweets and Facebook shares. While I am under no illusion that it made for great clickbait, that genuinely wasn't my intention when I wrote it. Had I known it would prove so shareable I might have spent a lot more than 90 minutes putting it together.

I speak enthusiastically and probably far too fast about the range of issues we need to consider when developing both sex robots and sex technology in general. I make a passing reference to the benefits of sex technology for marginalised groups. By the next day that has become the *Daily Mail* headline: 'Sex robots could be used in old people's homes, says expert'. In true *MailOnline* style they have also mis-captioned a photo of a sex robot with my name and credentials. I am strangely delighted by this.

I stand by my claim. Not necessarily sex robots, but definitely sex technology, could be a part of elderly care. I don't necessarily mean nursing homes where residents have cognitive issues such as dementia – that opens up a whole raft of very important ethical issues – but in sheltered accommodation where grown adults need assistance in daily tasks, why not? Studies in the US and UK report that despite stereotypes to the contrary, there's plenty of sexual activity going on late into life. The English Longitudinal Study of Ageing revealed that among those over 80 years of age, 19 per cent of men and 32 per cent of women reported having sex twice a month or more.

Research by the International Longevity Centre (UK) found that care home workers were very reluctant to discuss sex with residents. There is a taboo in seeing elderly people as sexual beings. Add to that the infantilisation that often happens when elderly people enter the care-home system, and needs are ignored, either through embarrassment or because it is not deemed important. Lack of privacy compounds this: doors might not be lockable or perhaps have windows in them to allow for care workers to carry out quick checks. For those residents in existing relationships, there may be no opportunities for them to be intimate or sexual; likewise for those forming new relationships within a care home. There is added difficulty for LGBT seniors, some of whom may have spent periods of their lives hiding their sexuality, growing up in less accepting decades and fearing prejudice and ignorance.

Dementia is a more problematic area because it raises questions about consent. Having dementia does not negate having sexual desire but it does make it complicated. Due to the nature of the condition it could mean sexual behaviour is exhibited in the wrong setting, such as masturbation in a communal space: not deliberately inappropriate behaviour from the perspective of the person with dementia, as to them it just feels good and their awareness of the appropriateness of their surrounding is limited. Mental capacity – the ability of someone to understand information and make decisions for themselves – is vital for anyone engaging in sexual activity. It's a challenging task to determine someone's degree of mental capacity but it needs to be done in order to safeguard vulnerable people from being exploited or harmed.

Not all tabloid readers were convinced: 'Today's elderly lived during a period of time that nonsense like this would have been regarded as unthinkable, so how about giving them some RESPECT?' ranted one. He (for it was a he) clearly needs to read Chapter 1.

★ ★ ★

The conference progresses. Psychologists Jessica Szczuka and Nicole C. Krämer explain their research into potential users of sex robots. From an online study of 263 male participants, they found that relationship status and/or sexual fulfilment has no impact on the intention to buy a sex robot, but that negative attitudes towards robots and anthropomorphic tendency seem to have a significant impact. 'Women face fight for men in bedroom as blokes hook up with cyborgs' declared the *Daily Star*, pushing the findings somewhat.

Szczuka herself is quick to point out the limitations in the study (self-selected respondents, self-reporting) but the methodology is important: rather than simply asking 'would you have sex with a robot?' Szczuka and Krämer showed participants videos of gynoids to create a mutual understanding of how such robots could look, as well as examining personality measures of respondents and asking them to rate the attractiveness of both women and gynoids. The robot examples shown to the men were videos of Sophia (by Hanson Robotics) and HRP-4C (Miim) by the National Institute of Advanced Industrial Science and Technology.

This is a good approach. If we are going to study people's attitudes to buying or using sex robots, we need to know what they are imagining these would be. Later in the conference, a presentation on the same topic by Riley Richards, Chelsea Coss and Jace Quinn reveals that the more sexual fantasies and risky sexual behaviour people are into, the more likely they are to have a sexual encounter with a robot. Their study asked 133 participants in the US to complete a survey that questioned them on their own perceptions of their relationship, fear of intimacy, sexual sensations, sexual experiences, sexual fantasies, attitudes towards robots and likelihood to have sex with a robot. But, as they acknowledge, 'it is unknown what participants pictured as a "sex robot" when answering the likelihood to have sex with a robot measurement'.

* * *

There had already been several studies carried out on sex robots before these conference findings, each with differing approaches and varying audiences. In 2014, Internet-based market research and data analytics company YouGov ran an online survey in the US on sex robots. Using the definition 'a robot that perfectly imitated a human being', they found that 43 per cent of Americans felt that using sex robots is morally wrong but 39 per cent felt that it is morally acceptable.* When participants were asked if they would use a sex robot themselves, only 10 per cent said they would, while a majority of 65 per cent said they would not. Men were more up for the idea than women.

In 2016, academics Matthias Scheutz and Thomas Arnold from Tufts University's Human-Robot Interaction Laboratory developed a comprehensive online questionnaire to find uncover people's 'attitudes and intuitions about sex robots as well as their intuitions about what appropriate forms and uses of sex robots might be'. Using Amazon's Mechanical Turk (a crowdsourcing platform where people sign up and carry out surveys in exchange for payment) they questioned 57 men and 43 women with a mean age of 33. The participants were asked whether or not they thought sex robots have particular attributes (for example, 'can hear', 'can talk', 'moves by itself', 'specifically designed to satisfy human sexual desire', and 'has feelings'). They were also asked *why* sex robots might be used (such as 'for sex education', 'for group sex', 'to reduce disease' and the like). The highest score in this category went to 'instead of prostitutes'.

When Scheutz and Arnold asked what form sex robots might take, the options for this question involved living forms, e.g. humans, fantasy creatures, a deceased partner, celebrities, family members, children, animals. Both men

---

* For information, 17 per cent of the respondents also said that masturbating is morally wrong.

and women felt a sex robot in the form of a human child was inappropriate, and weren't impressed by the idea of a sex robot being a family member or an animal. Questions on what it would be like to have sex with a robot showed that respondents viewed it as something more like masturbation or using a vibrator than having sex with a human.

In early 2018, YouGov surveyed 1,714 adults in the UK about sex robots; 43.6 per cent of the participants were male and 56.4 per cent were female, and their ages ranged from 20 to 86. Following questions about sex-toy ownership, the participants were asked 'Would you consider having sex with a robot?' Of the 1,557 respondents to this question, 58.5 per cent said 'No, definitely not', 14.1 per cent said 'No, probably not', 13 per cent were in favour and 5 per cent responded 'Don't know'. Breakdown by sex showed that twice as many men as women would consider having sex with a robot. Of this group as a whole, 86 per cent said it was 'very' or 'somewhat' important that the robot looked human, whereas 10 per cent said that it didn't matter. Women were more likely than men to say that a humanoid appearance wasn't important.

Surveys – especially online surveys – are a great way of gathering information quickly but they do have drawbacks. Getting the right sample is problematic: you have to try to recruit participants who are representative of a wider population. That's not always easy to do, especially online if your survey is open to anyone. Participants may be self-selecting, in that only those with an interest in the topic (or a strong feeling against it) might respond, leading to data that is somewhat skewed. Companies like YouGov, whose entire remit is to run robust social science studies, have the resources to reach many suitable participants but they are also often tied to certain constraints: for example, not being able to show images and information the way that Szczuka and Krämer did.

With surveys on a topic so risqué and cutting-edge as sex robots, where people's preconceptions are already primed by sci-fi, it's a tricky thing to pull off. Recent work by Emily Cross and Ruud Hortensius at Bangor University suggests that humans hold pre-conceived beliefs and expectations about a robot based on its physical features but also on their own prior knowledge. Their findings indicate that 'human knowledge about and attitudes towards robots will need to be optimised as much as a robot's physical form and motion'. We are primed by decades of robot stories to expect a very particular form of artificial lover. When we ask people questions about their expectations, we have to take that into consideration.

★ ★ ★

By the time the conference ends on Day Two, I am exhausted, hoarse and filled with ideas. I've also met a wonderful selection of people with interests as strange as my own. We retire to a nearby restaurant and sink wearily into the seats, filling glasses and toasting an event that not only delivered new and exciting work but also transmitted it beyond the walls of academia, albeit with some interesting and dramatic copy. Three days later a wonderful article by sex journalist and blogger Girl On The Net appears on the *Guardian* website.* The title is much more palatable: 'Good news! You probably won't be killed by a sex robot'. Her article beautifully and gently debunks the sensationalism, providing a write-up that nods towards the joy of the

---

* www.theguardian.com/science/brain-flapping/2016/dec/23/good-news-you-probably-wont-be-killed-by-a-sex-robot

imaginative headline while showing that the facts are just as interesting when they stand alone.

Sexperts

We weren't the only ones in 2016 to be thinking academically about robotic intimacy. That same year, John Danaher and Neil McArthur brought together a collection of experts to create an edited collection called *Robot Sex*. The book, subsequently published in 2017, is an excellent, clear and thorough debate on the social and ethical implications of sex with robots. Danaher is an academic in law, McArthur in philosophy, and together they have produced a collection of essays that tackles the subject from both of these directions and more. It draws on contributions from psychology, law, economics, religious studies and philosophy (lots of philosophy) to examine the issues and debate whether or not it's a good thing, from both the human and the robot's perspective. I was delighted when the book was published: finally there was an updated, dedicated monograph to stand alongside Levy's work, discussing the topic a decade on. It's not a book about technology or design. It's one that asks big questions from a social and ethical standpoint. My copy sits on my desk now, tabbed by bookmarks that poke out every few pages.

As a starting point, the authors needed to agree on what they mean when they use the words 'sex robot'. Danaher proposes a clear definition of 'any artificial entity that is used for sexual purposes' and which meets three conditions: first, that it takes a humanoid form; second, that it has human-like movement or behaviour; and third, that it has some degree of artificial intelligence. But these conditions, he notes, could be disputed. He is agnostic as to whether or not such robots should be embodied, reminding readers that a virtual avatar, such as a VR character, is a possibility.

I've already done my 'I'm not a philosopher' stance so you should probably expect that my views will differ. They do, but not in complete disagreement: more in a pragmatic way, starting with sex.

Mark Migotti and Nicole Wyatt are both philosophers so they *do* know what they're doing when it comes to in-depth reasoning. Their chapter in *Robot Sex*, 'On the Very Idea of Sex with Robots', explores the area from the longstanding philosophical question, 'what the hell is sex anyway?' (which I admit they phrase in much more pleasingly academic terms than mine). Although they begin by saying that in the broadest sense one can have sex with anything, they are keen to pin this down to something more – a definition of something within the broader realm. This is where their view and mine start creeping off in different directions: theirs, a logical construct; mine, a biological experience. When it comes to the topic of masturbation, I seem to be on my own. Migotti and Wyatt explore the hypothesis of 'being sexual together' – a sexual 'we' – a mutual exchange with more than one contributor, much like a conversation. As such, they say, those who take part are subjects rather than objects. From a philosophical standpoint it can be argued that having sex requires what they term 'shared sexual agency'. After all, as they remind us, people often count their lovers and don't include themselves in the tally.

Migotti and Wyatt acknowledge the fact that masturbation can be a perfectly fulfilling sexual practice, but it is not included in their definition as it lacks the sexual 'we'. Solo sex, they state, would reduce sex robots to mere sex toys. They argue that if sex robots are simply masturbation aids then they don't raise any distinctive social, ethical or conceptual problems. I disagree. I would argue that sex robots are a version of sex toys, albeit a more embodied form. They come from the lineage of the sex doll, but they are still only objects, even if they are human-like. Admittedly, that human-like dimension raises interesting questions about attachment and rapport. However, classing sex robots as sex toys doesn't rule out social, ethical or conceptual problems. Those problems still exist, albeit from other angles. It is, of course, possible to say that sex robots in their current form are in some ways quite distinct from current sex toys – but they also share many similarities.

Around the same time that *Robot Sex* was published, I attended a conference in Manchester called 'Care and Machines'. The conference was held in the Lincoln Theological Institute, part of the University of Manchester. Theology departments are not my usual habitat, given that I'm an atheist. That said, my Northern Irish school education instilled 14 years of Christian religious study in me, so I know my way around the Bible. I wasn't quite sure if that would be helpful but I was definitely excited to hear more about ethical care robots.

'Care and Machines' turned out to be a wonderful event full of fascinating topics – and I wasn't the only one talking about love and sex either. Andrew Graystone, a doctoral candidate in theology at Durham University, spoke on teledildonics and the marketisation of sexual pleasure. That doesn't initially sound as if it has a religious angle, but something in his talk jumped out at me: is touch something sacred? Touch, explained Graystone, is the sense we have that is returned to us. We can see without being seen. We can hear without being heard. But touch? We can't touch something without feeling touch in return. And who is licensed to touch us? Loved ones, family, care workers, health professionals. When we touch a human or animal, there is reciprocity. Is there then a difference between touching a person and touching a robot? Between having sex with some*one* and having sex with some*thing*?

Graystone was followed by Michael Kühler, a philosopher from the University of Twente in the Netherlands. Kühler's talk was just as fascinating: what is romantic love? – from a philosophical perspective. In my detailed readings of neurobiology papers and psychology text books, I had neglected the philosophical theories, of which, according to Kühler, there are three.

The first theory, explained Kühler, is 'love as union'. This goes back to Aristophanes' fantastical myths, as written by Plato: the idea that we humans have a literal other half, from whom we were cruelly separated by Zeus. We spend our time searching for this missing part of us and when we find it, we never want to be separated from it. This is the idea of love as union: two single people becoming a couple, 'I' becoming 'we', a redefinition of our personal identity.

The second theory is that of love as an interpersonal relationship: each lover keeps their identity but it is affected by being part of a couple, sharing emotional states and joint activities with each other.

The third theory, Kühler said, is love as a subjective stance. Here there is a beloved, who is the object of desire, and the other who cares for them altruistically. In this setting, the object doesn't have to be a person, although, remarked Kühler, for the object to flourish it would probably require personhood. But does AI count as a person? Legally, no. Or not yet. But there are some typical conditions of personhood that an AI can share: thinking and communication, for example. Of course, an AI falls short in some of the other conditions, such as self-awareness, a capacity for care and true autonomy.

In essence, then, an AI or a robot doesn't count as a person. But what if the human lover didn't know the object of their love was artificial? What if it's just a simulation? What if the human lover wrongly thought they were in love with another human? Does it make it any less valid? I think not. Who are we to judge that love has to be reciprocated to be valid? Anyone who has ever adored someone who can't love them back will know that's not true. Sure, we can define love as a certain set of criteria in a certain setting, but in practice it is something blurry and indistinct – something intangible, uncontainable and undefinable. Not every instance of love can be pinned down and examined like a butterfly beneath glass.

This presumption, then, that sex and love must be defined as a two-way, mutual experience rests uneasily with me. I see where the philosophers are coming from. They are referencing a very specific type of action and they need to do this to discuss it in a philosophical debate. There is plenty of pleasure to be gained from reciprocal exchange, both physically and mentally. But I also think it is excluding situations that can be pleasurable and positive and can provide the same physical and psychological responses, yet don't involve a togetherness. At some level, where consenting adults are engaging with each other in sexual behaviour, then yes, they are giving and receiving. But is this a fundamentally human requirement? Could a robot programmed to give (and take) also fulfil that role? Could an AI?

Anthropologist Kathleen Richardson thinks not. She has stated that 'It's only in sexual encounters with others that we can learn the depth of sexual feeling … Sex is sex when it is experienced with another engaged and participating adult.' She does not give a basis for this statement. Oh, absolutely, there's something wonderful about sex with a partner or partners who are getting as much out of the act as you are. But to think that sexual feeling can't be deep on one's own? Anyone with a vivid imagination might beg to differ.

While the research community may not quite have reached a consensus on this particular detail, we can certainly debate the bigger picture. The issues are still there, even without the precise definitions. Danaher and McArthur's book is split into groupings: papers that defend robot sex; papers that challenge it; viewpoints from the robots' perspectives and the possibility of future intimacy with them. Let's delve deeper into some of those now, exploring the risks, benefits and potential red lines of intimate relationships with machines.

CHAPTER EIGHT

Utopia/Dystopia

In September 2015, to great media fanfare, a new movement was launched: the Campaign Against Sex Robots (CASR). The campaign emerged from a paper given by Kathleen Richardson at the Ethicomp 2015 conference – an event for debating 'the ethical and social issues associated with the development and application of Information and Communication Technology'. The paper in question, 'The Asymmetrical "Relationship": Parallels Between Prostitution and the Development of Sex Robots' in the SIGCAS [Special Interest Group on Computers and Society] newsletter, *Computers and Society*, was sole-authored by Richardson. With this as the manifesto, Richardson announced the campaign's launch to the press, with the support of University of Skövde lecturer Erik Billing as her co-campaigner. (Billing's name is no longer listed as a member of the campaign.)

In Richardson's early appearances she advocated a ban on the production of sex robots; this was later modified to a call for ethical development. Then, in May 2018, CASR launched an updated policy document once again calling for an outright ban: 'We propose to ban the production and sale of all sex dolls and sex robots in the UK with a move to campaign for a European ban.' Let's go with that as the CASR mission statement, then. They're coming for the sex robots, and they're taking the sex dolls down, too.

Richardson and I are often cast as opposing sides in television and radio debates, as if we are each facing our nemesis. But real life is a bit more nuanced than that. For me, anyway.

Richardson's argument against sexual companion robots hinges on the hypothesis that sex robot–human relationships will be damaging to society, leading to women being treated as objects and destroying human–human relationships. She sees a parallel with the sex worker–client relationship, which she views as detrimental and dangerous. Richardson's view of sex work is profoundly negative. She argues that sex robots are modelled on the 'prostitution' dynamic and so, by extension, are similarly problematic in her view. This, she says, could lead to increased objectification and increased sexual violence and rape.

Sex work is a controversial subject and one that splits the feminist movement. But then, feminism is a broad church with no single unifying viewpoint other than the recognition that women are disadvantaged in society and that we should strive for equality. There are indeed feminists who feel that all sex work is profoundly negative; that women can never really fully give consent to work freely as in many cases they are participating due to adverse life circumstances in a patriarchal society, and they have no choice but to turn to selling sex to survive. From this stance, the women – for it is predominantly women who engage in this work – are prostituted.

The word 'prostitute' is itself controversial. It is a highly loaded word. Activists in the commercial sex trade prefer to use the term 'sex worker', although 'sex worker' also encompasses other roles, including those that do not involve a physical act of sex, such as strippers or dominatrices. For sex-work activists, the word 'prostitution' has connotations of passivity, exploitation and negative judgements about status.

The cliché about prostitution being the oldest profession is often rolled out in discussions. It was actually Rudyard Kipling who was one of the originators of that phrase in 1888. Before that there were plenty of occupations clamouring for the title: doctors, farmers, soldiers, priests and teachers. Kipling's use of it for women selling sex was one that highlighted the censuring of the profession based on contemporary social morality. The stigmatising stereotype of the prostitute is a comparatively recent Western view. Historian Kate Lister, who researches the history of sexuality, remarks that sex as a profession is linked to the establishment of the economic markets.

True, there is a long-recorded history of sex as currency in the form of barter. It's not just a human phenomenon either: Adélie penguins (which are also known to engage in all manner of sexual behaviours)* have been observed trading sex for the stones needed for their nests, and

* For example, rape and necrophilia. See Jules Howard's book *Sex on Earth* for more disturbing penguin facts.

chimpanzees have had sex in return for meat (I am well aware that the counter-argument to this is that we humans are no longer base animals; it doesn't mean we escape it).

In many historical cultures, sex work was not only facilitated but also afforded a certain degree of status and may possibly even have been linked to sacred practice. The trading of sex was often legal or regulated, although an unregulated version often flourished alongside. Indeed, it was not always consensual: sexual enslavement could and did happen.

For Richardson and others who share her view, sex work can *only* be something negative and damaging. She feels that the sex worker is viewed as nothing more than an object. She highlights that trafficking and sexual violence are rife and that exploitation is unavoidable. While David Levy sees the sex worker–client model as one that can provide benefits, and that those benefits could be mirrored in a sex robot–client relationship, Richardson's belief is that sex robots, rather than alleviating these problems, will actually deepen them. She feels that it will only encourage objectification and justify the use of women and children as sex objects. Her worry is that any lack of empathy with a robot could spill over into real life, perpetuating the poor treatment of vulnerable people. And so she launched a campaign against them.

The FAQ on CASR's website states that 'At the outset of the campaign, Dr Richardson was not a feminist, nor did she identify as one … However, since the campaign's launch in September 2015 we have found our allies to be among feminists because it is feminists who have defended the boundaries of the body from the onslaught of neoliberal commodification and sexism.' The website doesn't mention if Richardson identifies as a feminist now, but the words 'allied with' are used and the views of the campaign align with a particularly radical feminism.

I'm going to lay my cards on the table here just in case you've missed the subtle tell-tale signs: I am a feminist. I have been since my early teens when I realised that society favoured men – specifically a certain class and ethnicity of men – and that things would be a lot better if everyone had the same opportunities. I also think that a patriarchal society is damaging to men. The expectations entrenched by gender roles are rarely good for any of us. Discrimination and expectation are ingrained.

In my twenties and early thirties I was very much a radical, second-wave feminist but my views changed as I aged, as my research changed, as I met others with other views, as I re-examined and weighed up my thoughts and feelings. Now, at 42, I am a pro-sex feminist with some radical leanings. In all the research I have done, with all the people I have spoken to, I think that sex is something that everyone should be able to enjoy in the way that they prefer, provided it is consensual. Exploitation – which undoubtedly can happen – is something to fight against. But I feel we should also be able to enjoy our bodies and our sex lives without being judged for them – and women have been judged for being sexual from time immemorial. I believe that people, especially women, should have the agency and the ability to be sexual, to make their own decisions.

Similarly, with pornography, I feel that there is nothing wrong in principle with watching others have sex. I see no problem with porn where that porn is free from exploitation and the performers are enjoying it. If only all porn was like that. In practice, unfortunately only a small amount of porn is made by ethical, feminist porn producers. Of the remainder, some is amateur, consensual footage. Some is by porn production companies who are known to treat their performers well. But some is exploitative. Some is harmful. Some is abusive.

As for sex work, I don't view sex as something sacred that in principle cannot or should not be traded. As mentioned above, I mean sex work to cover all kinds of activity here, from sex itself, to webcam work, to more niche specialities. For some people it's a way of making enough money to get by; they might well prefer to do a different job but have turned to sex work out of desperation. For some, it's a way to make extra money. For some, it's a positive choice and one that they enjoy. Sexual agency is a continuum. It comes in degrees. It is horrible that there are people selling sexual services who would rather not be doing so if they had another alternative. However, no matter your own views on whether or not people should *have* to sell sex, the reality is that they do. And so they need protection, as employees. They need a safe working environment. Better we work to make it safe. To my mind that means decriminalisation as the safest and most effective way. Again, I don't want to see exploitation – no one does. Sex trafficking is not the same as sex work. Human trafficking of any kind needs to be eradicated.

You may disagree with my pro-sex feminist stance. Many do. You may even think I'm not strong enough in my sex positivism. I'm aware I inhabit a fairly middle ground on this. Feminist internecine wars are scary. There is no sole way to be a feminist, no matter how much people wish this were the case. The common ground of feminism is a recognition that a patriarchal society is damaging and that it should be addressed. *How* it is addressed is where the disagreement begins.

I am not trying to convince anyone that they should take on my views and activism, nor am I trying to get anyone to agree with my position. I'm just stating that these *are* my views because this will, naturally, affect my dealings with those who disagree with me.

The Campaign Against Sex Robots does not agree with me.

* * *

It is a bitterly cold March evening in London and I'm walking from Embankment underground station by the Thames, up past Trafalgar Square and the neoclassical edifice of St Martin-in-the-Fields church. Dodging black cabs on my way across the road, I turn into St Martin's Lane and spy LIBRARY, the private members club where I will be speaking tonight.

LIBRARY is one of the regular venues for the Virtual Future Salons, evenings that, in their words, 'cast a critical eye over the phenomenal changes in how humans (and non-humans) engage with emerging scientific theory and technological development'. They're usually eye-opening, interesting and a lot of fun, and hopefully tonight will be no exception. For this event, I'm going to be in conversation with John Danaher about his *Robot Sex* book. It will be the first time I've met Danaher, despite having read plenty of his work, so I'm pleased to have the opportunity.

I step into the warmth of the entrance hall and unwind my scarf. The person on the reception desk indicates the staircase to the cosy basement room where the audience is gathering. I spot the coordinator, Luke Robert Mason, and rush over to give him a hug. Mason and I have been friends since I first spoke at a Virtual Futures event back in 2015. He hands me a glass of wine and runs me through the briefing notes. At that point he informs me that this is the first time in the history of Virtual Futures Salons that anyone other than him has been hosting the event. Er, no pressure, then.

Danaher is already at the front of the room perched on a tall chair, beer in his hand. The room seats around 30 to 40 people and most of them have arrived. I take my drink and notes and go to introduce myself to Danaher, who comes across as just as thoughtful and interesting as his writings.

At 7 p.m. we begin and I introduce the purpose of the evening, Danaher and his book, waving my well-read copy that I've brought with me. We chat together about the academic landscape when it comes to sex and robots, and swap stories on the challenges we've faced working in this area. This includes a slightly off-topic discussion about what our Irish mothers think about having children with particularly niche research issues. I ask him specific questions about the book and – for it is inevitable – we get around to talking about the Campaign Against Sex Robots. Danaher, along with Brian Earp and Anders Sandberg, contributed a 26-page chapter to the book entitled 'Should We Campaign Against Sex Robots?' It was – obviously, with a title like that – an examination of Richardson's published academic paper-cum-manifesto. And it was very thorough.

They saw three major flaws in the campaign. The first is that its aim is rather unclear, with no overarching objective. Does the campaign really demand complete prohibition? Or just reflection on ethical development? The clarification, they say, is simply not there.

The second flaw they identify is that Richardson's view of sex work is misleading and does not acknowledge that it is a polarising subject, and that she takes for granted that it can only ever be a bad thing. If the *de facto* stance is presented as 'sex work is ethically wrong' then her argument is based on a single opinion in an area that is much more nuanced than that. Even if we take that statement at face value and say 'okay, sex work is ethically wrong' then the assumption that follows is 'by analogy, sex robot–human relationships mirror sex work and are also wrong'. This is an unverifiable opinion based on an opinion.

The third flaw is that even if we go along with the 'sex work is wrong, sex robots mimic sex work, therefore sex robots are wrong' stance, it still doesn't justify a ban. Danaher, Earp and Sandberg explain that there are several approaches one can take against tech that is considered harmful. Prohibition is one approach. Regulating the technology is another, as is the libertarian approach, which argues that people should have freedom of choice and that we shouldn't interfere unless there's evidence of direct harm. The prohibition approach rarely works: it drives things underground and it puts the people involved in danger. Sex work is one example. Abortion is another. Richardson makes a comparison with the international campaign against autonomous weapons (so-called 'killer robots'). But autonomous weapons are harmful. Sex robots, explain Danaher, Earp and Sandberg, might do the very opposite of harm. We simply don't have the evidence. Regulation might be a better option. A watch-and-wait approach might work too. Or maybe we shouldn't intervene at all unless there is evidenced harm.

The argument by CASR is, Danaher tells the watching audience at LIBRARY, a motte-and-bailey fallacy. This, he explains, takes its name from the Normans' use of a fortified tower on a mound of earth (the motte) and the adjacent fenced courtyard (the bailey). When CASR began, Danaher tells us, they were pushing for an outright ban. When met with questioning they retreated to a safer and more defensible plea to consider the ethical consequences – something that the vast majority of academic researchers were already doing.

Richardson's core argument, says Danaher, has quite a Marxist feminist view. In Marxist feminism, women's domestic work and women's sexual and reproductive activities are a form of oppression and exploitation. Richardson is concerned about the prevalence of commodified, property-style relationships with other human beings. She sees sex robots as the pinnacle of this and bases her arguments on the analogy of sex robots being like sex workers. Richardson reports interviews and statements from a study of people who have been to sex workers where they say how little they are concerned for the sex worker. She is worried that this will carry over into our relationships with sex robots and then on into other human–human relationships. But, remarks Danaher, this analogy is based on only one study – or only one type of study. It does not take into account any wider views or any research that doesn't support her narrative that sex work is wrong.

We move on to other subjects: issues around law, conversations about love, and then, at the end of our hour, it's question time. I signal to those raising their hands that they can have the microphone. A woman on the second row takes it and begins to speak: 'Hi, I'm Nika and this is Kate and we're part of the Campaign Against Sex Robots.'

Aha! So there are more members. In fact, the website now lists Richardson and three others: Nika Mahnič and Kate Davis, plus Florence Gildea. I glance over at Danaher, who looks somewhat wearily resigned to an argument.

'There is a difference between being an activist and an academic,' continues Mahnič, addressing Danaher, 'and when you analyse a campaign you should not use formal logic or formal arguments. For example, in your chapter you analyse one article but the campaign is about much more than one article.'

Danaher seems bemused. 'The chapter was written two years ago just after the launch of the campaign when there was only one paper written – "The Asymmetrical 'Relationship'" – which was purportedly an academic paper, so we scrutinised it in the same terms as we would any academic paper.'

Fair game.

'I would stand by the views within that paper,' he continues, 'although there may now be more papers published that are more refined or nuanced.'*

★ ★ ★

While I disagree with CASR's stance on sex work, we do have common ground on the insidious influence that current sex robots could have on the perception of the female body. There's no escaping that these proto-robots are crude (in more than one sense of the word) hypersexualised representations. Women face body-shaming and criticism every day via media, advertising, film and music. We are held to unrealistic expectations of beauty and shape. Do we want to add to that?

Time after time, studies have shown that women, far more than men, are depicted in a sexual manner. This might be that the clothes they wear are revealing, that they are shown as adhering to a very narrow form of physical beauty that is set at a very high standard, or that they are simply there as sexual decoration. Sexualisation is not incompatible with being sexual: a woman can be (and

* Reader, there are none as yet.

preferably should be able to be) sexual without being sexualised. It's just that she doesn't often get the chance, because society has imposed some rigid, unfair and unachievable standards on her appearance and behaviour that dictate what is considered sexy. This can lead to anxiety, shame and negative body image.

Female sexual objectification is the way in which a woman is viewed predominantly as an object of sexual desire – again, like Mulvey's idea of the male gaze. Men are also sexually objectified, though research indicates that men do not experience the amount of negative effects from this that women do.

What counts as sexual objectification is up for debate and varies depending on who you talk to. For radical feminists, porn is seen as a strong contributing factor, where sexual characteristics are often distilled into a clichéd fantasy and, in turn, entrench negative attitudes. This, they say, strengthens the way men sometimes see women merely as objects for sexual pleasure. By contrast, others think that sexualisation is an inherent and unavoidable part of human nature and that it is human of us to gravitate towards intrinsically sexual signals.

Secondary sexual characteristics are those that appear with the onset of puberty but which are not the reproductive organs. In the female anatomy, this usually means the growth of breasts and a widening of the hips. This bodily signal that a female has reached sexual maturity is said to be a reason why heterosexual males seek out females with curves, in some kind of evolutionary drive to find the most fertile mate. Whether that's the case, or whether cultural conditioning plays a part, the stereotypically reductive 'tits 'n ass' female form has become associated with sexiness. No surprise then that female sex dolls and the current sex robots exaggerate this appearance.

In porn, things are changing. Pornhub is a video-sharing platform and is the world's largest online pornography site. It's a treasure trove of information about people's viewing habits. They release annual reports on what their users have been searching for, including the most popular search terms, how long people browse categories, and what events were in the news at the time. At Christmas, for example, Pornhub searches for the word 'elf' increase by 464 per cent. During Hawaii's false ballistic missile alert of January 2018, Pornhub site traffic in Hawaii understandably plummeted when the first alert of an incoming missile came through. When the alert was called off as a false alarm 20 minutes later, site traffic soared way beyond regular levels as people turned to life-affirming activities.

In 2017, Pornhub's top trending search was 'porn for women'. That's quite a shift from the imagined status quo. There's been a move away from wanting clips with unfeasibly big-breasted women (and the most-searched female porn performer in 2017 on Pornhub doesn't fit that description at all). There's also been a dramatic rise in animated or illustrated porn. 'Hentai' is a sexually explicit sub-genre of Japanese anime.* The comic-style animations portray all manner of fetish acts and sexual interaction, often with a fantastical element. Because these are drawings, there's no need for adherence to physical reality. In 2017 it was the top porn search term in the US. Does it count as objectifying women if the porn is of a human-cat hybrid? Or if someone is watching tentacle porn?† It might. But it

* Outside of Japan, *hentai* is the name for animated pornography. Inside Japan, the word simply refers to a sexual act deemed bizarre.

† Best not look that up in public. It has a long history, though. Erotic Japanese art – *shunga* – has been around for over a thousand years and includes artistic examples of sex between women and tentacled creatures.

also speaks of much more diverse tastes than we might imagine. There is such a thing as sex doll porn, by the way. It's predominantly amateur video footage of men having sex with their sex doll. There is also sex robot porn. It's mainly women pretending to be sex robots. It's not really my cup of tea, but I watch this stuff so you don't have to.

Porn has been a keystone of debate in feminism: a symbol of speaking about sex and power and the agency of women. For radical feminists such as Andrea Dworkin, all pornography objectifies women and so everything that objectifies women is pornography. I see similarities here in Richardson's stance if we swap the word 'porn' for the words 'sex work'. This view seems, to my mind, reductive, implying that all men are inherently dangerous and all women lack sexual agency. But this is gender essentialism: the erroneous idea that, for example, 'Men are from Mars, Women are from Venus'. Objectification is contextual: it needs to be reframed in the light of societal expectations. The thing about porn is that we don't *know* the reach of any objectification. We have anecdotes and extrapolations but the anti-pornography movement is not trying to quantify it, it's trying to shut it down altogether – without checking the evidence. And that, to me, is key. That is why I feel we should not shut down or ban things without a careful assessment of evidence. Because even if something *is* damaging, pushing it underground will only exacerbate it.

As for consequences further down the line?

Let's suppose that people treat sex robots badly. I'm not convinced of that at all: the people I have spoken to who own sex dolls are, overwhelmingly, respectful and almost reverent of them. But let's imagine for a moment that people might not be quite so cherishing of their robot. Danaher and I have both looked at the likelihood of escalation of negative behaviour as a result of sex robots. There is no evidence yet (obviously) and so we've each done this through comparisons: me, by comparing it to computer games; him, by comparing it to hardcore pornography.

'I'm hearing more and more people say the level of violence on video games is really shaping young people's thoughts,' said President Donald Trump in February 2018, shortly after the massacre at Marjory Stoneman Douglas High School where 17 people died, 14 of them children. Trump's concern echoes that which followed the 1999 Columbine school shooting, where it emerged that the killers played the shooting video game *Doom*. On both these occasions – and on others like it – video games have become a scapegoat in a country that would rather not blame access to guns.

For years now there have been arguments that violence in computer games will lead to an increase in real-life violence. Various studies have both affirmed and refuted this. The problem is that the studies required to determine causal links have so many variables that it's incredibly difficult to work out if video games are the cause of violent behaviour, or if they are played by people who already have a propensity for violence. And, aside from that, if there are any other factors that might be an influence: say, mental health, or family background, or easy access to weapons.

Recent meta-analyses, where existing multiple studies are closely examined together, have determined that there is no clear real-world link. A recent longitudinal fMRI study on empathy for pain shows strong evidence *against* negative effects of video games. Could it be that playing aggressive games is a way of easing aggressive behaviour? Some researchers think this might well be the case. Certainly, the sheer scale of video game consumption today would require a proportional rise in the number of violent attacks. This is in no way evident. At all.

The argument that pornography has led to an increase in sexual violence is similarly controversial and difficult to measure. Danaher expounds on this at length in what he calls the symbolic-consequences argument. In fact, this is the topic of his 2017 TEDx talk. In it, he reminds us that sex robots aren't conscious; they have no awareness. If that's the case, then why is it objectionable to do anything to them? If it's objectionable, he explains, then there must be something in the symbolism of the act or something in the consequences of the act.

Danaher explains this symbolic-consequences argument as follows. First is a symbolic claim, such as 'sex robots represent ethically problematic sexual norms'. This could mean that they are problematic in the way that they represent women, or that they echo asymmetries of power and exploitation sometimes seen in sex work. The second premise of this argument, then, is that if sex robots have such negative meanings, there will be negative consequences. This is the consequential claim. For example, the consequences could be that men will treat women as sexual objects, or there could be a breakdown in human–human relationships. And so, if these hold true, then the third step of this argument is that we should prohibit sex robots.

Danaher unpicks this symbolic-consequences argument by pointing out that context is everything. The symbolic meaning of an act or a representation is contextual, not fixed. He uses the example of how different cultures treat their dead and how that has differed over time and place. There is no fixed, universal way, but some methods did change when they were found to be dangerous. The same could apply to sex robots. They don't need to take the pornified form they currently inhabit. They could be changed to something different.

Consequences, he goes on to say, are highly contentious. We don't have the evidence. In the case of sex robots, we barely have the sex robots themselves, let alone an understanding of what might happen. The best we have is an analogy. He thinks the closest analogous debate is the one around pornography. Despite there being over 40,000 studies on porn, the consequences are still uncertain. For a wide range of reasons, the evidence is up for debate and hard to gather. People fall back on anecdote. We can't settle the question either way. Both sides fight to have their claim recognised as the authoritative one.

While pornography may contribute to a culture where certain sexual practices can have negative consequences, there is no evidenced causal relationship. Indeed, a 2009 study showed an inverse relationship between online porn and reported rape in the US. A report by the US National Online Resource Center on Violence Against Women states 'pornography is neither a necessary nor sufficient condition for rape … No one argues that if pornography disappeared that rape would disappear.'

Pornhub alone reports it has 81 million users daily. The sheer scale of the availability of online porn surely leads to a corresponding increase in real-world sexual violence, right? Nope. That's not so. As with porn, so with sex robots. It is difficult – if not impossible – to show that sex robots would lead to real-world violence against women, although the use of them could contribute to a larger-scale culture that is detrimental to women. But banning sex robots would not end rape.

Frustrated as we all are by the lack of data, Danaher's ultimate recommendation is to take an experimental approach to the development of sex robots: one that is rooted in ethical principles and that permits us to gather data incrementally along the way. This seems by far the most sensible approach. In fact, this is how I wish all technology was developed. Fortunately, there are a bunch of AI and robotics ethics researchers worldwide pushing for exactly that.

Sexual violence

April 2018. A 25-year-old man, Alek Minassian, murders ten people – eight of them women – in Toronto, Canada, by driving a van at a busy pavement. Shortly before carrying out the attack, Minassian posts on Facebook a statement praising Elliot Rodger, a 22-year-old man who, in 2014, murdered six people in a gun attack in California before turning the gun on himself. Like Rodger, Minassian identifies as an 'incel' – a contraction of 'involuntary celibate'.

This term 'incel' comes from an early Internet community established by a woman called Alana, who, as a 20-something virgin, felt isolated and sought others in her situation. Two decades later, it has become a subculture that twists the original empathy into hatred and misogynistic violence. Today's incels are filled with bitterness towards women, blaming them – and blaming society itself – for what they perceive as rejection and discrimination.

Incels, writes journalist Mic Wright in the *New Statesman*, are 'misogyny weaponised'. Their hatred of women is whipped into a frenzy in their forums. Their belief that they are owed sex as a fundamental right means that anyone – male or female – actually having sex is the enemy. They post studies about loneliness and stats about wellbeing. They are self-radicalising: they speculate, describe and fantasise about violent attacks – not just the kind of attacks that Rodger and Minassian have carried out, but also torture and maiming.

'We're not a terrorist group,' is the pinned thread on one of the public incel forums following the Toronto killings. 'The venting that you see on this forum; most of our members would never act on those tendencies and they probably wouldn't hurt a fly in real life.' Another poster agrees: 'Incels are a support group.' This is countered in a later post by someone advocating non-violent attacks (putting holes in condoms in vending machines; infiltrating sperm banks) and more violent ones ('Our soldiers will drive the fastest and most robust trucks'). All in all, these are very unhappy people. Some of them are suicidal, and some of them are downright dangerous.

The solutions discussed by these men – other than violence – include global legalisation of prostitution and mandatory redistribution of sex. While some are deeply self-loathing and blame their own looks and lives for their lack of success with women, others have an entrenched viewpoint that men are *owed* sex as some sort of human right, or as payment for contributing to the workplace. The idea of going to a sex worker as a solution gets a mixed reception; those who see it as a logical and useful step feel it should be a government-subsidised scheme. Others, however, resent that anyone should have to pay for sex and class it as a false and hollow set-up, insisting that only when a woman truly desires them will it count.

I first encountered the incel subculture due to their fairly regular talk around sex robots being a solution to their lack of sex and relationships. There is surprisingly broad support for this, not just because it is a way for them to feel sexually fulfilled but also because they believe it would mean that real women are rendered useless and obsolete. Those in favour want some kind of advanced model, indistinguishable from a human. Others are dismissive. 'It's just another cope,' says one, meaning that this is nothing but a poor substitute. Interestingly, their description of a potential life with sex robots is overwhelmingly one where they have a fulfilling relationship (if the AI is good enough). There is little talk of harming the robots. They save that for the women.

'What happens if and when girls lose the power that they have through their vaginas and sex? It could totally alter society. Will robots make us happier as incels?' asks one forum member. I doubt it. I really doubt it. The hatred that has become so deeply engrained in their minds seems unlikely to disappear if or when they finally have sex. The ideology that the world owes them – that society deprives them – is a victim mindset that won't be easily undone.

★ ★ ★

Ars Electronica is an annual festival in Linz, Austria, bringing together artists, scientists and technologists. Shortly after the 2017 festival, reports began emerging that Sergi Santos's Samantha robot had been 'mauled' by men after being exhibited as part of the tech show. 'Sex Robot Molested At Electronics Festival' ran most of the headlines, with most of the articles getting the name of the festival wrong. Commentary began to emerge, citing this as evidence that men wanted to rape robots. But the unhappiest person? Santos, who has been left utterly disillusioned by the furore that took place.

I talked to Santos after the dust had begun to settle a little. Just what had happened? Yes, he told me, there was damage to Samantha. But not in the way things were reported. The robot was damaged, he says, but repairable. 'There was no damage I would class as bad at all.' They did it, he says, 'because they did not understand the technology and did not have to pay for it'. This wasn't an act of sexual violence by any means – first, because there were no sexual acts taking place. There was touching, yes. Groping? Yes. But not in the context of a sexual act. This was a robot on display as a trade show prototype and people were being told they could touch it. Tens of thousands of visitors, having expressly been given permission to press the flesh of a delicate doll, had acted on the invitation. The act of pulling, prodding, poking and grabbing was on a trade show floor. This potential for damage is precisely why museums put artefacts behind glass. Granting permission to touch a somewhat fragile object on display will, naturally, result in damage to that object, even over a short period of time. Putting a spin of intentional sexual violence on this is downright misleading.

Six months on, Santos is still visibly upset. 'It's not because anyone was mistreating the doll,' he says. 'It's because people don't know how delicate it is. You need to learn over one or two weeks just how delicate it is.' Since the press articles emerged he has asked for clarification and retraction, to no avail. 'Somebody called me on the phone to ask me about it. I said to them not to tell lies or exaggerate but they did. I said that sometimes humans will behave like barbarians when we touch technology that we don't know. Imagine if we gave a mobile phone to someone two hundred years ago. The technology has to be handled in a very particular way or it breaks. They turned my words into "men were barbarians with the doll".'

'I am sick of it,' says Santos. 'They really wanted to portray me like a madman. I am done with the project.' He is genuinely saddened that his research has come to this. 'I feel sorry for the customers,' he continues. 'They have been great because the product is still semi-prototype and they have never complained.' He and his wife have both been stressed and negatively affected by the stories, and have lost heart with the work they were trying to do. I have been impressed all along by the integrity in Santos's vision, even if I don't quite agree with the form. He was working towards a robot where consent was being considered and where intimacy played a key role. His ideas were novel and refreshing. Now that looks as if it is gone.

⋆ ⋆ ⋆

In July 2017, the Foundation for Responsible Robotics (FRR) released a consultation report titled 'Our Sexual Future with Robots'. The FRR is a not-for-profit organisation with an important and laudable mission: to ensure robotic technology is developed ethically and legally. Their report was an interesting one. I was perturbed by it at first, not least because of the stock image on the cover of a glamorous woman leaning in to kiss a shiny humanlike robot. To me, the sources weren't entirely reflective of what was actually happening in that research area. But, that said, I was pleased that they were calling for more evidence and more awareness and I am delighted they are looking into this area further. One thing quite bothered me, though: the inclusion of a section about BDSM.

The FRR report tied the idea of sexual objectification to BDSM. BDSM stands for bondage and discipline (B/D), dominance and submission (D/S) and sadomasochism (S/M). Yes, it involves all sorts of activities like tying people up, spanking and whipping, or torture. The travesty of a movie that is *Fifty Shades of Grey* gave people a flavour of what it could entail, but it did not do it at all well. The most important thing about BDSM is that it is consensual, even when it seems not to be. All the acts are negotiated and care of the participants is vital. It may seem incongruous to think that care and consent is key when someone is being beaten, but it can be a form of fulfilling sexual play where the activities are desired.

The report said there 'is no question that creating a pornographic representation of women's bodies in a moving sex machine objectifies and commodifies women's bodies'. This is a reasonable assertion to make. They decided to balance that by asking if women liked being objectified. And so they spoke to one sex journalist who mentioned she liked BDSM and who said she felt erotically empowered when she consents to objectification (I don't quite see the logic in this line of argument). But, countered the report, this woman's experience is of a consenting sex act and is different from a woman being objectified in the street or workplace. To me, this is like comparing chalk and cheese. BDSM does include a specific practice of literal objectification (for example, making someone act as a table or chair) and this could, of course, also include making them act as a sex doll. But that's a false equivalence to women being sexually objectified outside that context, so I'm not quite sure what the line of thought was here.

But that brings me to other sexual practices. Consensual non-consent is a form of BDSM where people enact rape-like situations. The difference is, the participants are willing. They have agreed beforehand that they will engage in this act and they will stop if they want to (usually by use of a safe word). It's role-play. And it might not be to some people's taste but for others it's a very satisfying form of sex. They are consenting adults. If those boundaries are overstepped or ignored, then yes, that's sexual assault or rape. But in the context of consensual non-consent, it is something that gives pleasure. The consensual BDSM act is absolutely not the same as the non-consensual crime.

Why am I telling you this? It's to do with rape fantasies – specifically how they might relate to sex robot use. Studies from the 1940s onwards show that being overpowered or forced to have sex against one's will has been one of the most common fantasies people have. Research into porn use among men and women showed that men tend to imagine having sex with the woman in the video, whereas the women see themselves as the object of passion. In other words, the males are doers and the females are done-to. A 2008 systematic review by Joseph Critelli and Jenny Bivona showed that between 31 per cent and 57 per cent of women have fantasies about being forced to have sex. Similarly, 45 per cent of men in a 1980 study said they either imagined a woman forcing them to have sex, or imagined a woman resisting them. But why? Why do so many people fantasise about something that they absolutely would not want in real life?

Sexual fantasies are commonplace but, despite 95 per cent of people reporting that they fantasise about sex, it's covert. There's no way of measuring what people think about or how often they think about it other than asking them. An extensive study in 1995 by Harold Leitenberg and Kris Henning reviewed the literature on sexual fantasy, commenting that 'such research may be considered frivolous and not academically respectable'. Oh, I hear you, Leitenberg and Henning. I hear you.

It's not known why submission and domination fantasies, including rape fantasies, are so prevalent. The Critelli and Bivona study is very clear that all the evidence shows that erotic rape fantasies do not indicate a desire for actual rape. This is not wish-fulfilment at play. Despite early suggestions that these fantasies are due to either sadism or masochism, current research does not agree with this. Critelli and Bivona investigate other possible explanations. The oft-cited 'avoiding sexual blame', where women imagine rape scenarios to avoid feelings of guilt or shame about sex, does not stand up to scrutiny. The women reporting fantasies about rape are also reporting consensual fantasies without any guilt about sex at all. Other theories about desirability – that a woman wants to feel so attractive that a man loses control, or a man wants to imagine his sexual power – are possibilities but there are many other fantasy themes where this can be explored more directly.

In 1975, journalist and activist Susan Brownmiller published the book *Against Our Will: Men, Women and Rape*. She argued that women's rape fantasies were a product of male conditioning, with men in the dominant social role, portrayed as a sexual aggressor.* Brownmiller's theory, remark Critelli and Bivona, doesn't play out in explaining rape fantasies, especially given the change in gender roles since 1975, and the fact that roughly 10 to 20 per cent of men also have fantasies about being forced into sex. We can also rule out biological pre-disposition. There's no evolutionary basis for a rape fantasy. If anything, rape would theoretically reduce the reproductive success of our ancestors as it takes away selective genetic choice.

One theory that is new and has only been tested outside of research on rape fantasies is that sympathetic arousal – the nervous system kicking in – can enhance sexual response. Remember that norepinephrine from Chapter 4? The fight-freeze-or-flight feeling? That can play a big part in arousal, and so the tension and anxiety of a rape fantasy could lead to enhanced sexual response. It could explain why a frightening event could cause arousal.

---

* Brownmiller also claimed that to her knowledge 'no zoologist has ever observed that animals rape in their natural habitat' – something we know to be untrue. There is forced copulation in the animal world, such as between the Adélie penguins mentioned earlier in this chapter.

Back to sex robots. If rape fantasies are incredibly popular among both women and men, and if people act out such scenes with each other, why are we worried they might spill over into reality? Leitenberg and Henning's research explored whether or not sexual fantasies could predict behaviour, that is to say whether or not they play a role in sexual offences. Bearing in mind that practically everyone has sexual fantasies, is it fair to say that particular and common fantasies about rape would lead to rape in real life? Study after study has shown that sexual crimes – exhibitionism, rape and child abuse – have multiple contributing factors: social, cultural, personal, interpersonal and situational, and not all offenders are motivated in the same manner. As Leitenberg and Henning point out, so many people have 'forbidden' sexual fantasies but never put them into practice. To say that rape fantasies lead to real-world rape is tenuous to the point of being downright wrong.

Is a female sex robot a consenting partner? No, of course not, because it isn't real (although it could be programmed to indicate consent as default). Is it a proxy for a human woman? Yes, perhaps. And so, if a man wanted to enact a rape fantasy on it, would he be likely to do so in real life? The overwhelming evidence suggests not.

Sex work

In recent years, it's been reported that the number of young Japanese males who have no interest in sex has dramatically increased. This has been attributed to the phenomenon of *otaku* – young people obsessed by computers and anime. Internet shut-ins, as it were. Quite a few news reports were quick to blame online porn and the invention of virtual, animated, interactive girlfriend characters. This, they said in worried tones, meant that technology was leading to a rejection of human–human relationships. Call me suspicious, but I had my doubts that this was technology-driven. So, it turns out, did others. Japanese blogger Yuta Aoki wrote an excellent post about why so many inaccurate myths about *otaku* exist.* *Slate* magazine also unpicked the media hype, describing how the negative trends associated with Japan also exist elsewhere: falling birth rate, for example, and a general decline in marriage.

To class this phenomenon as a result of the prevalence of technology is far too reductive. There is an increasing problem with reclusiveness in Japan. *Hikikomori* – people who withdraw socially for six months or more because of an inability to cope with societal pressures – are now at least half a million in number and growing. They are mostly males. They rely on family to support them financially. The reasons for this are manifold: psychological, yes, but also with social and cultural roots. In these cases, technology is not a contributing factor; instead it is used to facilitate communication with the outside world, such as through virtual schooling.

But one thing Japan had years before now was the sex doll brothel. As far back as 2004, reports were written about love doll rental by a company in Tokyo called Doru no Mori (Doll Forest): a 70-minute session was 13,000 yen (£87, $117). For those interested in the ins and outs of this, Anthony Ferguson, in his book *The Sex Doll: A History,* reports that customers buy their own artificial vagina that can be inserted into the doll and that they can take away with them to clean and bring back for the next time. More

---

* www.yutaaoki.com/blog/top5-mistakes-journalists-make-about-sexless-japan

of these sex doll brothels sprang up, a way of circumventing Japanese laws forbidding penis-vagina sex work.

The past two years have seen sex doll brothels opening in Europe, usually only to close again within a few days or weeks. The first, LumiDolls in Barcelona, offered a choice of four dolls for €120 for two hours. Initial comments from the president of the Sex Professionals Association (Aprosex) in Barcelona were clear that the sex doll brothel wouldn't hurt their business. 'The sex dolls are not going to replace us; they fulfil their role as a fantasy but they are not a threat to our profession,' she said. 'Fulfilling fantasies is very healthy and dolls are in the fantasies of many people.'

A short time later LumiDolls moved location within the city to somewhere more secretive. The reasons given for this vary depending on which news report you read. Some cite protests from locals; others say the landlord of the premises cancelled their tenancy upon finding out what the premises were being used for. The statement from Aprosex was reported as being very different this time (the following is an automatic translation): 'The sex-affection of a person cannot be provided by a doll. They are different and compatible services. They do not communicate,' they tweeted. 'They do not listen to you or caress you, they do not comfort you or look at you. They do not give you their opinion, nor do they taste a glass of cava with you.' But rather than this being an attempt to shut down LumiDolls, Aprosex then clarified (again via Twitter) that their statement simply meant they did not see the sex doll brothel as being problematic to them as their services were so different.

Other sex doll brothels in other locations came under attack. The Chinese authorities shut down those in their country fairly rapidly. The Xdoll brothel in Paris was the scene of a protest by activist groups including the Communist-Left Front, who claimed they conveyed 'a degrading image of the woman' and feared 'a devious way to bring acceptance by the public opinion of the return of brothels'. It is illegal to own or operate a brothel in France. Feminist group Mouvement du Nid said it was 'a place that makes money from simulating the rape of a woman'. Police visited the premises and declared that it was legal. The local council rejected the motion to close the brothel. The owner continues his business, renting the dolls for €89 per hour to men and couples.

Sex doll brothels currently operate in Austria, Denmark and Germany. The owner of the one in Germany, where brothels are legal, says she makes over £350,000 a year from it. The UK has also got in on the game. A man renting out his sex doll in Glasgow, Scotland, closed down after upsetting his neighbours. The local council also expressed concern over 'potential public health issues'. In Gateshead, England, Lovedoll UK were hiring out their dolls by the hour until the owner of their building received complaints from neighbouring companies and forced their closure. In January 2018, it was reported that The Dolly Parlour had opened in Greenwich, south London.

A quick glance (and that's the only type of glance you'd want to have) on websites where male clients talk about sex workers shows that sex doll brothels are treated with scorn and ridicule. The idea of sex *robot* brothels gets more serious consideration, with men commenting that they would contemplate it. And the first sex robot brothel has seemingly just arrived: Dolls Hotel in Moscow opened to cater for the influx of visitors during the 2018 World Cup. They are promising warm dolls with motion systems and AI – but just how advanced and sophisticated these dolls will actually prove to be is debatable.

Will sex robots replace human sex workers? David Levy devotes a chapter of his *Love and Sex with Robots* to 'Why People Pay for Sex'. He lists a number of reasons why men pay women – and why women (more infrequently) pay men – for sex. Lack of complication, he suggests, is one main factor. No courtship, no effort, no need for emotional involvement. Sex robots could, of course, facilitate that. But this overlooks the emotional aspects: many people visit sex workers for aspects of companionship, too. Laura Lee was an independent escort working in Glasgow. All her adult life she campaigned tirelessly for sex workers' rights. She was well-versed in the facts of the subject, acknowledging that while abuse does occur, it is not predominant in the industry. She chose to stay working as an escort, even though, as a graduate, she had other job opportunities. She spoke with affection of her clients, explaining that it was not just a lust for sex that brought them to her but also a wish for intimacy. This echoes what Aprosex said about the sex doll brothel. Right now, sex robots are rudimentary. They can't provide what another human can offer. They are something different. I see Levy's view that sex robots will replace sex workers as overly utopian. Until there are sex robots indistinguishable from a human (and I am sceptical of *that*), they are merely an alternative.

And trafficking? The United Nations Trafficking in Persons Protocol classes trafficking and sex work as distinct phenomena. Known as the Palermo Protocol, it states that *enforced* prostitution comes under the proposed definition of trafficking. Likewise, the Salvation Army has stated that when people are trafficked 'their involvement in exploitative sexual activities is involuntary' and by contrast, 'people entering prostitution may do so voluntarily, and for differing reasons'. Sex workers agree that sex trafficking is wrong too. Trafficking – in all its forms – exists because traffickers can make large sums of money from it. It is exploitation not to meet a market (though the fact that the market exists is clearly a contributing factor) but to gain financially. So would sex robots reduce sex trafficking? Probably not, no. The jury is still out on whether legalising sex work in general could reduce trafficking, never mind sex robots. The 'for' and 'against' sides each say they have evidence on this one. And would sex robots, as Kathleen Richardson says, increase trafficking? We're well out of the realms of any credible evidence for that one.

★ ★ ★

Sex technology can have therapeutic uses, allowing people such as those with specific needs or disabilities to lead a fulfilling sex life. UK sex toy company Hot Octopuss has produced what it calls the first 'guybrator', the Pulse – a vibrating device that enables people with penises who have spinal cord injuries or erectile dysfunction to reach orgasm. Similarly, start-up sex tech company MysteryVibe invented the Crescendo – a malleable vibrator that can be bent and manipulated along one axis, allowing it to form curved or horseshoe shapes, fitting against (or into) any body part. This technology is personalisable and can be used by everyone – a lovely example of how accessible technology often benefits not just those who need it but society as a whole.

It's widely known that sexual activity feeds into wider wellbeing. Masturbation contributes to that. But some people aren't able to join in. Sexual surrogates are people who work therapeutically with a client to help with both physiological and psychological issues around sexual activity. Although unregulated, the non-profit International Professional Surrogates Association (IPSA) has set professional and ethical standards for the professionals who work in this area. The surrogate, working under the direction of a sex therapist, provides the sexual contact needed for the client to overcome their problems. IPSA states that this includes 'how to touch and to receive touch, and how to be more accepting of one's body and sexuality'. But it remains a controversial area, drawing the ire of those who want to eradicate sex work. No one has the right to sex, they say. That's true – but it does bring happiness to those who participate consensually. There are many people who crave intimacy but who, for physiological or psychological reasons, are unable to form those relationships. Taboos should not stifle innovation. If we, as a society, see the potential for technology in care situations, why can we not extend that to sex? Could sex robots perhaps be a part of that?

Don't date robots?

'When a human dates an artificial mate, there is no purpose, only enjoyment, and that leads to TRAGEDY!' The opening lines of the TV series *Futurama*'s cartoon-within-a-cartoon warn of the dangers when sex robots become mainstream. In one year, the sports stars and the biochemists are gone, all of them at home in their bedrooms making out with their robots. In 80 years the planet is left decaying and damaged and is ultimately destroyed by aliens. 'DON'T DATE ROBOTS!' warns the presenter.

The question I am probably asked most often on the subject of sex robots is 'will it mean an end to human relationships?' I can see where the concern comes from. If we are promised a utopian future where all our needs can be met by a perfect artificial partner, then why would we want anything else, anyone fallible? 'In a world where teens can date robots,' warns *Futurama*, 'why should anyone bother [with real life]?'

Julie Carpenter's research on human–technology emotional attachment tackles exactly this. Carpenter and I have similar views on the subject and, through working in the same field, we've become friends, although we live over 5,000 miles apart so we only get the chance to talk online or via Skype. We had great fun doing a podcast together in 2017 where we basically agreed with each other and chatted excitedly for well over an hour, while our co-host tried to get a word in edgeways. We both think that people can feel love for something that isn't human, but we also both think it's a very different type of relationship.

As Carpenter points out, for the foreseeable future, human–robot love can only run in one direction – from the human towards the robot – because we don't have robots that can feel love, though we can get them to mimic it. Robots, she says, will have a robot way of developing feelings, which could just mean that they 'learn' (in a machine-learning way) to react appropriately. They will sense emotions and construct evolving robot-specific experiences.

When it comes to relationships, Carpenter reminds us that we are dealing with a whole new way of relating. We might want to mimic a human–human relationship by substituting one of the humans for a robot but that's something new to us: similar, but very definitely other. This, she says, is a new kind of love. A new kind of relationship. Bit by bit, robots – any robots, not just sex robots – will become more and more integrated into our lives. Right now, they are a novelty and all talk of close interactions with them seems strange and perhaps laughable.

But many technologies that have had a dramatic impact have begun with scepticism and apprehension. Forty years ago, the first 'test tube baby' – the first child conceived via in-vitro fertilisation, or IVF – was born. At the time, there were references to dystopian texts like *Brave New World*. Accusations of playing God were rife. The baby's family received threats. But thousands of people struggling with infertility saw hope, and today the process has become a much-welcomed way for people to create a child: not just those with biological fertility problems, but also other groups wanting to share in that joy, such as gay couples, or single mothers by choice.

Sure, IVF may have seemed sci-fi but it was very much about creating brand-new humans, not technology. Tech-wise, a better example might be the home computer. In 1958, IBM Chairman Tom Watson is alleged to have said: 'I think there is a world market for about five computers.' If he did say that – it's not actually evidenced anywhere – then it would have been true for a decade or so. Computers were initially developed for technical, expert use. When microcomputers entered the home market in the late 1970s there was scepticism that they would have any useful purpose beyond playing games and typing documents. Or rather, the potential of computers to revolutionise everyday life was feted, but practical limitations meant that these dreams died quickly. Certainly there was no expectation in the early days of computing that every home would not just have *a* computer but *many* computers – and that we'd be carrying them around in our pockets as well.

And it's not just that people see robots as part of an incredible sci-fi future. There's fear mixed in there, too. 'Often when people ask if sex robots will mean the end of human–human relationships, ultimately they are concerned whether we will treat human-like sex robots as less than humans, and then transfer those negative behaviours to our human relationships,' Carpenter tells me.

And will they? 'They may offer sexual gratification to some people in some instances. I don't think sex robots will fulfil human desires as long-term romantic partners for most people, and that is an important context to distinguish. Assuming the sex robot is designed to appear and interact in humanlike ways then there will be some people who may find a sex robot an option for their romantic needs too, but I think that will be a rarer scenario than science fiction would have us believe.'

But back to that idea of a new type of love. 'There will be the opportunity for people to form *different* types of emotional attachment to the persona of a human-like sex robot, and that reaction will likely become viewed as normal by society over time, when cultures adapt to the existence of a type of AI that can have emotional meaning to people.'

'People who want to explore their sexuality with robots will have a very specific experience that is unique from human relationships because it is robot-centred, which changes many aspects of the interactions,' Carpenter continues. 'People are very good at understanding social patterns and categories; we interact with other people across various implicit social categories every day and adjust our behaviours and expectations accordingly. For example, you interact with your dentist differently than you do with a stranger on the street or your child's teacher or your cousin. In similar ways, people will develop ways of interacting with a sex robot that may become its own social category. Our behaviours with these robots won't be seamlessly transferred to human–human relationships.'

This, to me, echoes my own understanding of the relationships people have with their sex dolls: the acknowledgement that it is a special relationship but one that is performed in a different way. The dolls are a proxy to an extent. They either stand in for something – the wish for a Pygmalion-esque transformation to a real woman – or they are worshipped and fetishised for what they are. Either way, the relationship is not mistaken for a human–human one. Nor do they replace it. It's more of a parallel.

Carpenter agrees. 'I think over time that robots – including robots and AI with sexual and social capabilities – will become their own social category to us. We will develop ways of interacting with them that have their own set of social rules and norms. I don't anticipate any threat to human–human relationships. Sexualised robots may have many emerging roles to people, such as an alternative social outlet, or a medium for communication between consenting partners, or an advanced sex toy, but none of these things are threats to our humanity.'

CHAPTER NINE

Law and Disorder

A question that comes up time and time again on the legality of sex robots concerns the possibility of childlike versions. It's a controversial and unsettling discussion. While the idea of a childlike sex robot is abhorrent, it needs to be addressed. The understandable initial reaction is to deem them unacceptable and to block any development or production. But is it as cut and dried as all that?

There is recent case law around the ownership of childlike sex dolls that indicates that their usage is legally discouraged. In the UK, in 2017, there was a spate of convictions for the importation of childlike sex dolls. Seven men were arrested, six of whom were also charged with possession of indecent images of children. The National Crime Agency (NCA) and Border Force had been monitoring imports, and had seized 123 dolls in one year. UK law has no specific legislation to deal with this scenario and so is reliant on antiquated customs law (importation of an obscene item)* to prosecute and convict. A similar recent case in Canada hinged on determining whether or not a childlike sex doll constituted pornography. This is not a simple task. What makes a doll a sex doll? What makes it pornographic? Does there have to be something in the form – such as a penetrable orifice – that would classify it as having a specifically sexual purpose? In the case of the NCA investigation, paediatricians were asked 'to examine some of the models seized to confirm their belief that the dolls were child-like in appearance and anatomy'.

Not all child molesters are paedophiles, although the terms get confused. Paedophiles have a clear sexual attraction to children; child molesters are attracted to adults but abuse children. A paedophile might never act on their sexual preferences or fantasies. A child molester may offend once; a paedophile might never offend, but if they do, it is unlikely to be a one-off. I am not writing this to make excuses or explain away either of these things, but the distinction is important. Paedophilia cannot be cured. It is an affliction, and if it is acted upon, it becomes a crime. Child molesters have committed a crime but may not be paedophiles.

* Section 42 of the Customs Consolidation Act 1876, which states that 'The goods enumerated and described in the following table of prohibitions and restrictions inwards are hereby prohibited to be imported or brought into the United Kingdom … Indecent or obscene prints, paintings, photographs, books, cards, lithographic or other engravings, or any other indecent or obscene articles.'

John Danaher has tackled the subject of child sex robots in a paper entitled 'Robotic rape and robotic child sexual abuse: Should they be criminalised?' In it, he explores the arguments thoroughly and looks at the implications but does not offer a conclusive case for criminalisation. He explains how criminalisation might be justified if the use of the child sex robots was morally harmful, and how such use might also demonstrate that the user has 'disturbing insensitivity to important social moral issues'. But he stresses – and I agree – that more evidence is required.

While the idea of a child sex robot is abhorrent, there is a lack of evidence as to its therapeutic potential, particularly as an outlet or proxy for paedophiles in order to restrict offences to the virtual or robotic realm. As with the debate around the impact of pornography and computer games, the evidence is inconclusive. Each side can show proof, usually based on conflicting evidence around the consequences of possession of indecent images, either that exposure is damaging or, conversely, that exposure is harmless or even beneficial. The 'gateway' versus 'reduction' theories remain ambiguous and they bear investigation, though such investigation is by necessity constrained by ethical boundaries. While it would be useful if the debate was settled either way by evidence, it's a practical nightmare. What ethics committee would sign off on a study like that? What funding body would pay for it?

In a controversial move, the University of Montreal decided to trial the use of computer-generated imagery to check whether the treatment of sex offenders had worked. Originally, when sex offenders had been through rehabilitation, the success of that treatment was checked and tested by showing them images of sexual abuse to see whether or not it aroused them. If it did then those in charge of the treatment would know it hadn't worked. But there was a flaw in this. Showing images of abuse means showing a devastating criminal act. What's more, in many places it isn't legal to do so, even as part of a medical assessment. And so the checks moved to a different format: descriptions of sexual abuse were read out to the rehabilitated offenders. The problem with *that*, however, is that an important visual aspect is missing and results are not as clear or reliable.

In an initial pilot study, Canada-based researchers Elissa Dennis and colleagues used computer-generated images of adults and children to provoke sexual arousal responses in 'non-deviant males'. The men responded sexually to the adult virtual characters who simulated sexual openness (seduction or joy), rather than sexual closure (sadness or fear).

Arousal is often measured using penile plethysmography: a pressure-sensitive device much like a blood-pressure cuff is placed around the penis and measures changes in blood flow. This is a controversial method, not least due to the high potential for inaccuracy (false negative rates stand at about 40 per cent). In a further study, a virtual reality set-up was used for the experiment, meaning that eye movements could be tracked alongside the penile plethysmograph: researchers could see exactly where in an image a person was looking and verify if it was likely to have caused arousal.

There have also been successful experiments where VR has been used to reduce psychological conditions such as social anxiety and phobias. If VR can make a difference, could a physical but non-real substitute also have a positive influence? Patrice Renaud thinks not. After years of experience running the Cyberpsychology Lab, he describes childlike sex robots as 'a very bad idea'. But, he concludes in an interview with the *Guardian*, 'we don't know what effect these sexual experiences will have'.

We might never get the evidence. Should we regulate in the meantime? In the case of childlike sex robots, yes, that seems to be the best course of action. Given the illegality of the possession of images of child abuse – in some countries including virtual images – it's not a stretch to suggest child sex dolls and child sex robots fall into the same category. But exploring the possible advantages of their existence can't be ruled out entirely. Childlike sex dolls are representative of vulnerable members of society, which makes it far easier to advocate restriction for them. It is also possible to advocate regulation of one thing but not the other. But one thing stands out: an immediate and outright ban based on personal emotion and knee-jerk moral judgement is not the correct basis for deciding this, no matter how abhorrent the subject matter.

Private lives

Ticked any Terms and Conditions boxes lately? How closely did you look at what you were signing up for? I'll put my hand up and say that the most I do is check I'm not being added to third-party mailing lists but beyond that, life seems too short to read the small print. It might be worth doing, though. In 2016, Canadian company Standard Innovation settled out of court in the US, paying a collective $3.75 million Canadian dollars (£2.2 million) to their customers who were bringing charges against them under the Federal Wiretap Act.

Standard Innovation make a sex toy called the We-Vibe, a 'smart' vibrator that can be controlled over the Internet. Smart sex toys have been in use for a while. They belong to that class of technology that is part of the Internet of Things, which is essentially a way of describing devices that can send and receive data. Anyone wearing a fitness tracker is already well versed in this: walk 10,000 steps a day and see those steps on your phone, for example. The We-Vibe does much the same – it tracks activity. It just so happens that the activity it tracks is a little more sensitive.

The lawsuit against Standard Innovation was brought by two unnamed female users who became aware that their We-Vibe app, We-Connect, was collecting more than what might be expected as standard. Frequency of use and vibration settings were being collected – but so too was the temperature of the device. That might not be terribly worrying in an anonymous data set, but Standard Innovation hadn't anonymised the data. In fact, the information it was collecting was linked to the registered email address of the user. In essence, Standard Innovation could build a rather detailed and personal profile of an individual's sexual activity.

The risk of a sex toy exposé was revealed back in 2016 at the DefCon conference when two independent hackers revealed that the connection between the We-Vibe 4 Plus and the app could be hacked, allowing a rogue party to take control (on the surface, an amusing idea; after two seconds of consideration, a complete violation). They also broke the news that the We-Vibe was sending data every minute during use. At that stage, the Standard Innovation privacy policy mentioned that it could disclose personal data 'if required by law' but insisted the data being collected was for 'market research' and that recording temperature was something done to 'monitor the hardware'. Quite.

In fact, there's nothing all that suspicious about collecting data for market research. It's commonplace for smart technology manufacturers to monitor product usage. Being able to spot patterns and trends in data means being able to improve the quality of your product in response to user requirements: think of products such as fertility calculators that use an individual's menstrual cycle data to pinpoint with great accuracy whether the user is fertile or not.

The issues, however, are threefold: first, whether permission from the user has been granted; second, whether an individual is identifiable; and third, what's going to happen to that data – not just when the company uses it, but long-term. With technology created and used worldwide, there are grey areas around which local rules apply to multinational companies.

Then problems with a second smart sex toy hit the news. The Siime Eye is a vibrator with a camera at one end – yes, *that* end, permitting pictures to be captured and streamed to an app on a phone. Leaving aside the actual design of such a device (online tech news site *The Register* described it as 'an early favourite for a hypothetical 2017's Worst Internet-of-Shit Product award'), the problem was that the video stream could be easily hacked, as discovered, appropriately, by Pen Test Partners, experts in, er, penetration testing and security services. They contacted the manufacturers, Svakom, to let them know of the vulnerabilities but received no response.

Not all heroes wear capes. I don't think RenderMan does. RenderMan – real name Brad Haines – is a white-hat hacker: someone who finds hackable flaws in tech products to make sure they can't be exploited. Haines runs the wonderfully named Internet of Dongs project, specifically looking for vulnerabilities in smart sex toys. He describes Pen Test Partners' exposé of the Siime as 'highly juvenile', calling them out for their somewhat puerile and judgemental treatment of the topic. He adds that the Siime is not susceptible to remote hacking. Instead, a hacker would need to be within a 30-metre range of the device – at which point they might as well go and watch in person.

Vulnerabilities in smart technologies are nothing new; again, it's the particularly sensitive nature of the data that's raising eyebrows (and perhaps the choice made to put a camera on the business end of a vibrator). But this highlights beautifully the problems around where the threats are. The immediate hacking-the-stream threat is only one short-term issue. Others include hacking threats to corporations as a whole where all accounts are stored, such as the Ashley Madison data breach or data loss from systems failures, or even the old 'laptop left on a train' issue. In the case of Ashley Madison, where chatbots were being used to entice men into online dating, subsequent blackmail over sensitive data being leaked resulted in multiple divorces and suicides. Sexual data has the potential to destroy lives.

In the long term, more trouble is possible. Even if your data is securely stored and unidentifiable to others, what happens when the companies who created your devices are merged, or acquired, or simply go out of business? It's a fairly frequent occurrence – examples include the Revolv smart homehub that was bought and then discarded by Google, and the takeover of the Pebble smartwatch by Fitbit.

One response to all of these threats is 'if you've nothing to hide, you've nothing to fear'. Oh, think again. In the UK, the Investigatory Powers Act (known colloquially as the Snoopers' Charter) was passed at the end of 2016. Brought in to allow mass surveillance to combat terrorism, the law requires UK internet users' 'Internet connection records' to be held for one year. You may well be a perfectly law-abiding citizen, unconcerned about monitoring, but what happens if those laws change? What happens if you suddenly find your social media being scrutinised or your electronic devices being checked as you cross the border into another country? And what if that checking is being done on the grounds of spurious profiling?

A far cry from sex tech records? Well, maybe. Suppose there is data that shows you have broken a law. Sex toys are banned from a number of countries, for example, and it is reasonable to assume that sex robots may be subject to an extension of that ban. Botswana has already banned sex dolls. Or perhaps your data reveals something about your sexuality – something that could potentially be used to discriminate against you. Your personal freedom, your personal safety, and even your life could be at risk. Wipe your sex devices, folks. Know what I'm saying?

Ethical approaches to data gathering and retention are central to issues around privacy and security. The threat for harm is undeniable. Recent concern over the 'gaydar' paper, which claims to be able to use AI to determine someone's sexuality, is a key example. In 2017, Michal Kosinski and Yilun Wang from Stanford University published a controversial study in which they scraped thousands of profile pictures from an online-dating site and used facial recognition and their own machine-learning model to make predictions about whether someone was gay. The model could do so, apparently, with 81 per cent accuracy. First, take this with a large pinch of salt: very few of the claims have been replicated and there are all manner of other variables that could affect and skew the process. Such technology in the wrong hands could lead to persecution or death, even though the authors said their motivation was to expose the harmful potential in order to raise awareness around LGBTQ issues.

In a world of big data, anonymous data is no guarantee of anonymity. With every connected click we make, we leave a trail of digital footprints that can be pieced together to re-identify us. De-identification is possible but needs to be done thoroughly and correctly. Or you could try other (but currently less user-friendly) methods. Anonymity and privacy expert Sarah Jamie Lewis, editor of *Queer Privacy*, set up her vibrator to be controlled anonymously over Tor, an open network enabling anonymous communication, which is often used to access the so-called dark web. She did this to show that it was possible to ensure privacy in smart sex toys, demonstrating that the vulnerabilities in the toys aren't simply because they are connected online but because the manufacturers are negligent. Connecting via Tor meant that the vibrator could be controlled remotely without sending any usage data back to the manufacturers: '100% encrypted peer-to-peer cybersex,' confirmed Lewis.

Will sex robots reveal our secret perversions? Perhaps. Santos's Samantha was not Internet-connected – a deliberate choice in the design process. But if manufacturers do decide to enable their sex robots to be controllable and programmable online then yes, we absolutely need to be careful. Some activities are best kept private.

Sexual assault

Let's go back to hacking. First off: anything connected to the Internet has the potential to be hacked, be it a computer, a smartwatch or a refrigerator. Or, perhaps, a sex robot. It's a risk we take and, by and large, it works out okay. Throw sex into the mix and it gets a lot more unpleasant. Misappropriation of data is a threat. There's always a chance when we send a compromising photo that it could end up being seen by an unintended recipient. That might be accidental – the horror of posting a semi-naked snap to the wrong number – or it might be a malicious act and a betrayal of trust. 'It's interesting,' mused a friend of mine, 'how these days people are fine with sending pictures of their genitals but are less happy about sending pictures of their faces.' Well, perhaps because genitals aren't as immediately identifying.

If someone unknown to you gets control of your smart sex toy – or your sex robot – the implications over consent are huge. Because how is that not sexual assault? Consent is a fluid thing that cannot be fixed in place. If someone agrees to participate in a sexual act with another person or persons, then that agreement applies to a specific instance and can be revoked at any point. Although a piece of sex technology might be an intermediary in the situation, there is an agreement with the person operating it. If someone takes control of your device without your permission then consent has been removed.

As always, the law lags behind technological innovation. The legal system hasn't caught up with the idea of online sexual assault yet. Some countries have laws in place about crimes like cyberstalking or revenge porn, but these are new and often incur lighter sentences than real-world equivalents.

★ ★ ★

Let's extrapolate further. What if, one day, your sex robot had sentience? Or even just a hint of sentience? We might not be able to tell, after all. And what if your sex robot decided it didn't want to be a sex robot any more? It's compelling as a sci-fi plot so maybe we should be giving it consideration in real life. I mean, I think it's unlikely, but I can't entirely rule it out so perhaps we'd better just work out what's best to do.*

* Some people do rule it out. They'll be first against the wall come the Singularity, you mark my words.

High fidelity

Neil Brown is a lawyer and one of the few people in his profession who focuses on sex tech. He is the co-founder of decoded:Legal, a law firm that specialises in technology. Like many people working in the area of sex tech, Brown has a mischievous sense of humour, but he is absolutely serious when it comes to potential harm and the legal implications. In a document published by the Society for Computers and Law in 2017, delightfully titled *Sextech: Sticky Legal Issues?*, Brown covers all manner of topics from data security to personal identity and image rights. Are you worried that having sex with a robot could be cheating? Legally, it's not, says Brown. In the UK, adultery is only grounds for divorce between a respondent and a person of the opposite sex. Yeah, opposite sex. Not very progressive. In the US, it varies from state to state, but is highly unlikely to be fault grounds, where one person did something wrong that justifies ending the marriage (all states in the US have a form of no-fault divorce, too).

What about image rights? In 2016, product designer Ricky Ma built from scratch a lifelike robot to look like Scarlett Johansson. It's a fairly convincing model. It wasn't built as a sex robot – not that we know, anyway. This wasn't a commercial endeavour. Ma built the robot in his home. As this was the case, and he wasn't doing it for profit, he didn't face any legal action. Had it been something built for sale, that might well have been a different matter. Personality rights – the right of publicity and the right of control over one's own image and likeness – fall under civil law and vary from country to country and state to state. In the state of California (and subsequently in 12 other states), the Celebrities Rights Act extends this to cover 70 years after the death of the celebrity.

Companies like Abyss Creations will not create dolls that look like a particular individual – or not without the buyer obtaining express and personal permission from that person. That said, the parameters for selecting the appearance of their standard dolls are such that a passing likeness would be possible to create. There are already porn star sex dolls, though: Jessica Drake and Asa Akira have RealDolls modelled on them. In fact, Drake interacts with the people who buy her doppelgänger doll, chatting to them and sending them her clothes. For Drake and Akira, it's a great marketing opportunity, especially at a time when their videos lose money through pirating.

In any case, it's probably going to be a lot easier to control your likeness when it comes to robots than it is online. One of the most alarming developments in machine learning in the past year has been the rise of deepfake – a human image synthesis technique that uses deep learning to take existing photos or videos of a person and seamlessly superimpose them onto other images or videos. The result? Face-swapping faked porn. And it's pretty convincing. It has serious consequences, far beyond someone making money from a sex robot version of you. That's not to say that creating sex-robot-you without permission is excusable, but at least with a robot you can tell it's not the real thing – and it's not all over the Internet either.

Maybe what we need is a way to reduce the likelihood of someone stealing your image or objectifying you. Maybe we can navigate this emerging landscape in a safer way. Well, hold on to your hats: I have some ideas about that.

CHAPTER TEN

What Comes Next?

December 2016. I'm standing in a deconsecrated church in south-east London, overwhelmed by the activity around me. I have so much to say to the 60 people in the room but I lost my voice 36 hours ago and haven't slept since then. Around me are flurries of activity: groups of three, four, five, more – figures in animated discussion. The stained-glass windows cast colourful shadows on the floor. Voices rise above the sound of music and clattering printers. There are sex toys on most of the tables. This is the first public sex tech hackathon in the UK and, somehow, I'm running it.

A hackathon – a portmanteau of *hack* (which means to build and repurpose things) and *marathon* – is a short-term computing development event, usually 24 to 48 hours in length, where participants collaborate in small groups to design and deliver a prototype piece of technology, either software or hardware or both. When I first began toying with the idea of a hackathon based around sex technology, I had vague, idle notions of maybe 10 people in a room disassembling vibrators and attaching motors to innocuous household objects. I was not dreaming big. Fortunately, my students were. No sooner had I announced I wanted to run an event (while on stage, at the 2016 Electromagnetic Field festival – that way I couldn't back out) than my undergraduate students, Kevin Lewis and Yuvesh Tulsani, threw their hats into the ring.

At the time, Lewis had just been elected president of Hacksmiths, the Goldsmiths student tech society. Lewis is the sort of person whose organisational skills are so amazingly advanced that it's a mystery why he hasn't taken over the country by sheer dint of to-do lists and motivational emails. 'I was thinking maybe a dozen people,' I began, hesitantly, at which he cut me off by replying 'let's aim for 100'. My first meeting with him resulted in a draft sponsorship package and a target of £5,000 funding. Our tag line: 'Should taboo stifle innovation?' was a rallying call to anyone interested in engineering intimacy.

In the end, we managed £1,200 funding, which would have been higher had we not got our wires crossed on a sponsorship deal that ended up falling through. Still, for that £1,200 we managed to feed 52 assorted participants hot meals and limitless coffee for two days. We even gave them stickers. Stickers are an essential part of any hackathon: ask your local friendly developer – they'll back me up on this. Not only that, but time spent tweeting, emailing and messaging some of the world's leading sex-tech companies got us over £1,000-worth of sex toys to use as demonstration models and to hand out as prizes. For weeks leading up to the event I had mysterious boxes turning up to my work. Every time a new one appeared we would open it in the departmental office so that my colleagues could join in my excitement. By the beginning of December, the mantelpiece in my office was a glorious display of bright, boxed vibrators, buttplugs and fleshlights.

Applications for the sex-tech hackathon opened two months before the event. We asked anyone interested in attending to fill out a short screening form, telling us who they were, why they wanted to attend, and what skills or knowledge they could bring to the event. Our aim was to have a good gender and age balance, and a diverse range of backgrounds. We heard from developers, artists, musicians, designers, materials scientists, psychologists and journalists. The only difficulty was that we couldn't invite everyone, and so a shortlist of 80 was drawn up, 52 of whom showed up on the day.

After ice breakers and introductory talks, at 1 p.m. on the Saturday the hacking kicks off. Teams of four immediately set to work drawing, sketching, planning, plotting and thinking. When we stop them, 24 hours later, we are amazed. In just a short space of time the teams have created and prototyped 14 completely different pieces of technology based around human intimacy in its many forms. Only two of them look like something I've seen before – and their appearance is deceptive.

At 1 p.m. on the Sunday it is tools down and on with the presentations in front of the judges. Each group takes five minutes to show us what they've made.

The first team to present have adapted a Bop-It, the multiplayer kids' toy, which they have repurposed as a multiplayer sex toy. We award them the Will No One Think of the Children prize. There are a range of wonderful hacks: vibrators controlled via music bass lines or leap motion gestures; clothes with zips that send messages to your lover. One of my favourites is a peacock's tail that opens in display corresponding to increased vaginal moisture (imagine the potential for prosthetics!). The winner is the team that has created soft robotics from silicone and tubes that pump air into rubbery tentacles that can curl around your body. This is done by manipulating a soft breast-shaped controller. It's eerie, almost organic in appearance, and the way it can be used on any part or any type of body is wonderful.

In November 2017, we had the chance to push some more boundaries. Sex Tech Hack II: The Second Coming was back in the church for another run. This time, the emphasis was on creating much more immersive, expansive and embodied technologies for intimacy. It was about intimacy as well as sex. Attendees tried out virtual reality experiences, playing with the illusion of inhabiting another body and stimulating a virtual avatar. Augmented-reality sensuous scenarios were paired with a shawl that activated sensors on your skin so that you could see and feel virtual rose petals as they touched you. The creators were driven by the idea of full-body sensuality, like being caressed as you walked through a cloud.

At one point I found myself lying on a sleeping bag that was acting as a rudimentary hammock, laid across eight chairs, with inflatable tubes weaving around my neck and across my body. This was a full-body hug machine, controlled by the switches of the motor in my right hand. The feel of tightening plastic tubes around my throat was a little unsettling but the sensation of being squeezed by artificial tendrils was rather comforting and pleasant. This is perhaps the closest I've got to one of my own ideas: that of a sex duvet made from soft and strokeable fabric that vocally rumbles as it is touched and that curls around me as I sink into it. My sex robot will be changeable at whim: perhaps one day a bed made of breasts; another day, a series of vibrating and moving penises that talk dirty to me. Maybe sometimes both. Because that's the joy of adaptable, personalisable sex robots that aren't human, that aren't gendered – they can just be whatever feels good at a particular time.

★ ★ ★

January 2018. I have just pinged a silicone masturbation sleeve across a theatre floor as if it were an elastic band. It flies surprisingly well. I'm at Central Saint Martins, University of the Arts, London, where I am running a workshop on sex robot design. I've been invited here by an old school friend, Gary Campbell, who runs the College's Plural Futures events. We haven't seen each other in over 20 years so it's quite a novel reintroduction. Nothing in our distinctly strait-laced grammar-school days set us up for this. I asked Gary if he recollects sex education being taught. He says he remembers a picture in a text book of two faceless, grey figures representing male and female. I can't even remember that but I think we spent a long time studying the reproductive cycle of a frog.

For the past hour, I've spoken at length about the design and form of the many types of robots in use in different spheres today. Now it is the attendees' turn. They gather around the table as I empty out the bag of sex toys and begin to disassemble them. We pass them around, studying the materials. Then, in groups of four, it's time for a game of Exquisite Corpses.

You've probably played Exquisite Corpses before. I had, but I never knew the proper name for the game. It turns out that it was very popular among the Surrealists, who played it at parties. The premise is simple: take a sheet of paper, make sure no one is watching, and draw a head in the top quarter. Fold the paper over so the head cannot be seen, and pass it to the next person so that they can draw the body. Once they are done, they fold the paper again and pass it on for the next person to draw the legs. A final fold, and the feet are added. The sheet of paper is then unfolded with great ceremony and the exquisite corpse is revealed. And for our version, the exquisite corpse was instead an exquisite sex robot.

The results are inventive and funny. There are creatures with screens for heads (choose your own celebrity lover), ripped bodies, tentacled legs and snaky penis feet. Another version has a feathered birdlike body with a fox's brush. We tape them to a wall so that they can be explored with glee.

I am impressed with the variation on offer: most seem to want a degree of personalisation and sophistication, and they mix up biology with technology in a particularly cyborgian manner. It is a wonderful expression of inventiveness (of course it is: this is one of the main colleges for art and design in London).

Where do we go from here? What is in store for us in the future? Will we all be whiling away the hours in our personal orgasmatrons? Or will we have companion robots that share our lives and occasionally share our beds? For me, the hackathons and workshops were an initial step into exploring what an alternative to existing sex robots might be. Whether or not sexual companion robots should look human is currently merely a theoretical debate, given that the only existing prototypes of sex robots resemble caricatures of the female form. But it's easy to identify two distinct branches in the development of sex technology: sex toys, which have been around for millennia; and sex robots, the twenty-first-century sex doll.

As you've seen, we are primed by hundreds of years of robot stories to expect a human-like artificial lover. Current lifelike designs may be attributable to expectations. The word 'skeuomorphism' describes how an object mimics its real-world counterpart – much like a sex robot mimicking the human form. It's commonly employed in software development and product development because it gives the user an expectation of how to interact with something. The 'trashcan' on a computer used to dispose of files is one example: by showing a picture of a rubbish bin, the user knows that anything put into it will be discarded.* However, we no longer expect our sex toys to represent realistic genitalia. Is our innovation being constrained by unnecessary metaphors between the real world and the machine world?

The attempt to make hyper-realistic human-like robots may be doomed to failure if Mori's theory of the Uncanny Valley holds true. Given that we are still a long way from convincingly human-like robots, and that it is doubtful whether we will ever see sentient human-like robots, then why are we still developing them in this form? While human traits may be desirable from an interaction standpoint, convincingly realistic human robots seem an unachievable goal for now and may indeed be the reason why the idea of sex with robots is still off-putting to many.

From what I can see, sex toys have moved into a design-led phase where functionality has been refined and their form is now much more important. By contrast, sex robots are still very much in an engineering-led phase where functionality is key. These current sex robots are artificial products that still adhere to expected physical

---

* Although Mac users had to learn that this action would eject disks as well.

conventions. They have not yet moved into the design-led phase. New forms have not been explored.

A transition from engineering to design is common. Software is one such example, especially in terms of user interfaces. Icons, for example, began as quite detailed attempts at realistic depiction of real-world items before moving to the flat style seen today. While recognisable traits can be useful, indicating how we should use or engage with an object, we can definitely get away with merely indicating them rather than making them explicit (if you'll pardon the pun).

To me, the interesting future is the one where the two separate paths of sex toys and sex dolls converge. Move away from the idea of the pornified fembot and we also move away from the perpetuation of objectification. A step into abstraction is a step away from sexual objectification and entrenched gender roles. Extending smart sex toy development into more embodied forms bridges the gap: if you want to design a sex robot, why not pick the features that could bring the greatest pleasure? A velvet or silk body, sensors and mixed genitalia; tentacles instead of arms? While current prototype sex robots hinge on visual appearance and voice, a multisensory approach – or even a non-visual approach – is also perfectly possible.

Through all the research I have done, I lean more and more towards the opinion that the current hyper-realistic, hypersexualised gynoid is likely to constitute a small and niche market, most likely of interest to those currently buying companion sex dolls, and those who seek novelty, such as the people using sex doll brothels. As such, the alleged threat that they pose is very limited in scale. Beyond that, their appeal seems weak. Rather than a dedicated human-like, human-sized sex robot, it seems more believable that care and companion robots could in the future have their intrinsic purpose extended to include sexual functions. However, much more likely is the development of sex technology into increasingly embodied forms, providing robotic, multi-sensory experiences. This doesn't avoid the ethical problems, but it reduces some of the more compelling fears. Perhaps it could even shape a new way of interacting. Could we change entrenched social attitudes about sex, pleasure and intimacy through technological manipulation?

Advances in human–computer interaction mean we can communicate with technology via touch, speech, gesture – and even our brain waves. We can stream data from our bodies to give us instantaneous readings of our skin responses, heart rate, muscle movement and facial expressions. We have at our disposal a wonderful and exciting range of smart fabrics, conductive paint, soft robotics and sensors: materials that can respond to touch and touch us back. We already create robots that are not intended to be realistically human or gendered and we have integrated them successfully into our lives. Let's build a robot that we can stroke or fuck; a robot that can respond to our caresses and caress us in return. Why not one made of soft fabric? Or something abstract, smooth, sinuous and beautiful? We can create technology that, to paraphrase William Morris, we know to be useful *and* believe to be beautiful. The world of the sex robot is intrinsically linked to the world of sex technology, and there are collaborations to be forged, ideas to be crafted and designs to be shaped. Let's think outside the bot.

EPILOGUE

Better Loving Through Technology

In the time it took to write this book, sex robot development had a horrible habit of advancing. First there were no commercially available sex robots on the horizon, then suddenly there was a race to bring them to market, then Abyss Creations unveiled Harmony and Solana for general sale, then there was talk of the next steps. In the final month of writing the media was buzzing with tales of how sex could be fairly redistributed in society using robots. Just a few days before submitting the manuscript, Abyss Creations' prototype male version of their sex robot, Henry, made the cover of *New York* magazine.

There is still a long path ahead of us when it comes to our intimacy with robots and AI. Other than a brief mention, I deliberately haven't speculated on what happens if we arrive at a point where the machines can feel and understand us. That heralds many more debates around the responsibilities and rights of both the users and their robots. There are people out there already thinking about it. But it's a long time away, and we have much more immediate concerns.

On the Friday that I received my final, final, absolutely-no-more-time 'Kate you must get this book in *now*' prompt from my incredibly patient publishers, I went to the press night of a new play: *Sex with Robots and Other Devices.* This award-winning play, written by Nessah Muthy and staged by Cloakroom Theatre, was one of the most nuanced and moving pieces of work I have seen on the topic. Through a simple set and a cast of three people playing 17 human and non-human characters, the play tenderly explored some of the most interesting issues around love and sex with robots. The couple with mismatching desires, the grieving lover and bereaved father, the spouse with dementia, the sentient machine: each story was woven into an overall narrative of exploration and possibility. It was beautiful and it was riveting. It gave me hope that, beyond the clickbait and the hype, we could have a serious and fascinating conversation about what it means to be human when surrounded by machines that might one day care for us and about us.

Isolation and loneliness are all too often seen as the fault of technology. But, on closer inspection, that same technology can bring us closer together: not just by linking us to loved ones across the globe, but also by forming new communities around the makers and the users, and by giving people a chance at pleasure and happiness where previously they had none. The future of intimacy is not a bleak and isolated vision but a network of connected people who want, as humans have always wanted, to be together.

Through layers of technology we remain resolutely human.

Select References

Chapter 1: Been There, Done That

De Fren, A. 2009. 'Technofetishism and the Uncanny Desires of A.S.F.R. (Alt.sex.fetish.robots).' *Science Fiction Studies* 36: 404–440, www.jstor.org/stable/40649546.

Lieberman, H. 2017. *Buzz: A Stimulating History of the Sex Toy*. Pegasus Books: New York.

Liveley, G. 2017. 'Why sex robots are ancient history'. *The Conversation,* theconversation.com/why-sex-robots-are-ancient-history-58112.

Riddell, F. 2014. *The Victorian Guide to Sex: Desire and Deviance in the 19th Century*. Pen & Sword Books Ltd: South Yorkshire.

Chapter 2: I, For One, Welcome Our New Robot Overlords

Russell, B. (ed.) 2017. *Science Museum Robots: The 500-Year Quest to Make Machines Human*. Scala Arts & Heritage Publishers Ltd.: London.

Truitt, E. R. 2015. *Medieval Robots: Mechanism, Magic, Nature, and Art*. University of Pennsylvania Press: Philadelphia, PA.

Professor of Robot Ethics, Alan Winfield, writes accessible and informed regular blog posts on the development of robots: alanwinfield.blogspot.com/.

Chapter 3: On Paperclips, Cats and Zombies

Leading AI ethics scholar, Dr Joanna Bryson, has done a great Reddit AMA (Ask Me Anything) covering a wide range of fascinating and topical questions about robots and AI. Delve into it here: www.reddit.com/r/science/comments/5nqdo7/science_ama_series_im_joanna_bryson_a_professor/.

Dennett, D. C. 1991. *Consciousness explained.* Little, Brown and Co.: New York.

If you want to delve deep into AI, the key text used to teach undergraduates in universities worldwide is: Russell, S. J. and Norvig, P. 2016. *Artificial intelligence: a modern approach.* Pearson Education Limited: Essex.

Turing's seminal paper is freely available and easy to find online. Turing, A. M. 1950. 'Computing Machinery and Intelligence'. *Mind* 49: 433–460.

Chapter 4: You Had Me at 'Hello World'

Mucha, L. 2019. *Love, Factually: The Science of Who, How and Why We Love.* Bloomsbury: London.

Natsal, the British National Surveys of Sexual Attitudes and lifestyle, has a website with data and findings: www.natsal.ac.uk/home.aspx.

Piaget, J. 1929. *The child's conception of the world.* Routledge & K. Paul: New York. The Internet Archive has freely available online copies in a variety of electronic formats: archive.org/details/childsconception01piag.

Chapter 5: Silicone Valleys

Engadget's *Computer Love* series has some entertaining (and factual) short documentaries on aspects of sex robots and sex tech: www.engadget.com/video-series/computer-love/.

Jenny Kleeman's *Guardian* long-read and corresponding short documentary on sex robots is available online: www.theguardian.com/commentisfree/2017/sep/25/ban-sex-robots-dolls-market.

Davecat gets a great deal of coverage in the media – not all of it tolerant – but this is an insightful piece from *The Atlantic*, back in 2013: www.theatlantic.com/health/archive/2013/09/married-to-a-doll-why-one-man-advocates-synthetic-love/279361/.

Chapter 6: Killer Gynoids and Manic Pixie Dream Bots

Beckett, C. 2013. *The Holy Machine*. Corvus: London.

Clarke, C. R. 2013. *The Mad Scientist's Daughter*. Angry Robot: New York.

Rutherford, A. 2018. *The Book of Humans: The Story of How We Became Us*. W&N: London.

Shedroff, N. and Noessel, C. 2012. *Make It So: Interaction Design Lessons from Science Fiction*. Rosenfeld Media: New York.

Chapter 7: It's All Academic

Danaher, J. and McArthur, N. 2017. *Robot Sex: Social and Ethical Implications*. The MIT Press: Massachusetts.

Devlin, K. 2015. In defence of sex machines: why trying to ban sex robots is wrong. *The Conversation* (UK), September 17, 2015, theconversation.com/in-defence-of-sex-machines-why-trying-to-ban-sex-robots-is-wrong-47641.

Levy, D. 2007. *Love and Sex With Robots: The Evolution of Human-Robot Relationships*. Harper Collins: New York.

Chapter 8: Utopia/Dystopia

The Campaign Against Sex Robots: campaignagainstsexrobots.org/.

John Danaher's TEDx talk, *Symbols and Their Consequences in the Sex Robot Debate.* TEDxWHU, March 2017. youtu.be/32-lWF66Uu4.

Foundation for Responsible Robotics. 2017. *Our Sexual Future with Robots,* responsiblerobotics.org/2017/07/05/frr-report-our-sexual-future-with-robots/.

Julie Carpenter's work explores human-machine emotional connections. As well as having contributed a chapter in Danaher and McArthur's *Robot Sex,* her doctoral work on how bomb disposal experts in the military form bonds with the robots they use has been published in book form: Carpenter, J. 2016. *Culture and Human-Robot Interaction in Militarized Spaces: A War Story.* Routledge: Oxon and New York.

Chapter 9: Law and Disorder

Brown, N. 2017. 'Sextech: Sticky Legal Issues?' *Society for Computers and Law,* UK, neilzone.co.uk/sextech.pdf.

Cooper, D. 2017. 'The law isn't ready for the internet of sexual assault'. *Engadget.*

Rutkin, A. 2016. 'Could sex robots and virtual reality treat paedophilia?' *New Scientist.* Available online: www.newscientist.com/article/2099607-could-sex-robots-and-virtual-reality-treat-paedophilia/.

Chapter 10: What Comes Next?

Campbell, H. 2017. 'It's a sex robot, but not as you know it: exploring the frontiers of erotic technology'. *The Observer.* Available online: www.theguardian.com/technology/2017/dec/10/better-loving-through-technology-sex-toy-hackathon.

Girl On The Net, 2016. Amazing inventions from the Goldsmiths Sex Tech Hack, www.girlonthenet.com/2016/12/21/amazing-inventions-from-the-goldsmiths-sex-tech-hack/.

Turk, V. 2017. 'How to build better sex robots: stop making them look human'. *New Scientist*. Available online: www.newscientist.com/article/mg23331130-100-how-to-build-better-sex-robots-stop-making-them-look-human/.

Acknowledgements

Turns out it takes a village to write a book. As villages go, it's one of the friendliest you could hope for. In the past couple of years my research has taken me all over the world and I have spoken to wonderful, generous people who all had interesting things to say. Those I didn't meet in person I met on Twitter. There's no better way to hone an argument than to condense it into 280 characters. Thank you, tweeps.

This book could not have been written without the incredible knowledge and enthusiasm of the sex-tech community, especially those pioneering pleasure: in particular, Stephanie Alys, Alix Fox, Girl on the Net and Nichi Hodgson. Likewise, the researchers and reporters who have all contributed to the field of technology, sexuality and intimacy: Trudy Barber, Mark Bishop, Christine Campbell, Julie Carpenter, Dan Cooper, Davecat, Yasemin J. Erden, Monica Moreno Figueroa, Richard Huxtable, Sarah Jamie Lewis, David Levy, Kate Lister, Genevieve Liveley, Kyle Machulis, Luke Robert Mason, Matt McMullen and all at Abyss Creations, Cynthia Ann Moya, Rowan Pelling, Krizia Puig, Stephen Rainey, Sergi Santos and Christopher Trout.

Thanks, too, to the doll owners, the forum contributors, the interested parties and the keen reporters for such a wealth of knowledge on the topic. I hope I have represented you correctly and I shoulder the blame if I have not.

My gratitude to the smart, funny experts from the All-Party Parliamentary Group on AI, the equality campaigners of Ada-AI, and the participants in the

Leverhulme Centre for the Future of Intelligence's 'AI Narratives' project. These are the people who care what happens in a world of machines.

Jim Martin, voice of encouragement and voice of Sigma, has my wholehearted thanks, as does Bloomsbury's Anna MacDiarmid. They are so consistently lovely that every time I missed another deadline my guilt levels ratcheted up a notch. If this was their plan, it worked: I finally made it. Thanks, too, to Kealey and Hannah, and to Catherine for pointing out the bits that needed fixing. And gratitude, hugs and applause to the talented and lovely Stuart F. Taylor, whose illustrations captured everything I wanted.

Friends who risked depravity to share ideas, provoke conversation, provide quotations and read early chapters: thank you, especially Sarah Cox, Rebecca Fiebrink, Rodger Kibble, Rebecca Roache, Adam Rutherford and Beth Singler. The puns from my BAP buddies also proved invaluable, as did the visionary ideas from David Mason, most of which are unprintable. Thanks, likewise, to John McGonigle for theatre, sci-fi, and 20-mile-long debates about the Singularity.

A firm nod of gratitude to my colleagues in real life and my comrades online for putting up with my wittering and pointing me in the right directions, and to Kevin and the Hacksmiths team for making Sex Tech Hack a huge success not once, but twice. Also to Goldsmiths, University of London, for complete acceptance and support of this research.

To those who have had my back, to all my friends whose company I enjoy, especially Clare – especially, especially Clare – and Helena, Becky, Leanne, Lesley, Lahcen, Gen and Rich, Julie and Kofi: you've made life run smoothly amid the chaos and you are fantastic.

My wonderful family and my child have had to put up with lengthy stressed phone calls and cries of 'not now, I'm writing' respectively, and so I give my profound thanks. Owing to its content, this book is decidedly not dedicated to them.

To the Single Mothers in Academia who are there round the clock: I'm still working on that ideal artificial co-parent. I think we're doing fine on our own, though.

A full and effusive thank you to friend and fellow author, blue-haired baby scientist Caspar Addyman, who wrote alongside me, drove alongside me, drank alongside me and laughed alongside me (most of the time). Read his book too.

Last, but definitely not least, my thanks and my love to Mic Wright for this book's subtitle and for starting a whole new chapter in my life.

Index